I0140266

by **WAR DEPARTMENT**

75-MM TANK GUN M2

(MOUNTED IN LEE MEDIUM TANK M3)

FIELD MANUAL

May 4, 1942

©2013 Periscope Film LLC
All Rights Reserved
ISBN#978-1-937684-49-5
www.PeriscopeFilm.com

DISCLAIMER:

This document reproduces the text of a manual first
published by the Department of the Army,
Washington DC. All source material contained in the
reproduced document has been approved for public
release and unlimited distribution by an agency
of the U.S. Government. Any U.S. Government markings
in this reproduction that indicate limited distribution
or classified material have been superseded by
downgrading instructions that were promulgated by
an agency of the U.S. government after the original
publication of the document. No U.S. government
agency is associated with the publishing of this
reproduction.

©2013 Periscope Film LLC
All Rights Reserved
ISBN#978-1-937684-49-5
www.PeriscopeFilm.com

FM 23-95

BASIC FIELD MANUAL

⚜

75-MM TANK GUN M2

(MOUNTED IN MEDIUM TANK M3)

UNITED STATES
GOVERNMENT PRINTING OFFICE
WASHINGTON : 1942

WAR DEPARTMENT,
WASHINGTON, May 4, 1942.

FM 23–95, 75-mm Tank Gun M2 (Mounted in Medium Tank M3), is published for the information and guidance of all concerned.

[A. G. 062.11 (2–27–42).]

BY ORDER OF THE SECRETARY OF WAR:

G. C. MARSHALL,
Chief of Staff.

OFFICIAL:

J. A. ULIO,
Major General,
The Adjutant General.

DISTRIBUTION:

D 17 (6) ; R 17 (5) ; IBn 17 (3) ; IC 17 (10).

(For explanation of symbols see FM 21–6.)

TABLE OF CONTENTS

BASIC FIELD MANUAL

75-MM TANK GUN M2

(MOUNTED IN MEDIUM TANK M3)

CHAPTER 1

MECHANICAL TRAINING

Section I

CHARACTERISTICS AND DESCRIPTION

■ 1. CHARACTERISTICS.—The 75-mm tank gun M2 is a flat trajectory weapon of the field gun type. It is single shot with a drop type of breechblock automatically operated. It fires projectiles which weigh approximately 15 pounds. (See par. 2.)

■ 2. GENERAL DATA.—*a. Weights, dimensions, and ballistics.*— The following weights, measurements, and ballistic data are given for the information of all concerned:

Weight of 75-mm gun M2, complete_____pounds__	783
Length of bore_____calibers__	28.47
Length (muzzle to rear face of breech ring) _inches__	91.75
Caliber _____mm__	75
Type of breechblock_____verticle sliding__	
Muzzle velocity:	
AP shell_____feet per second__	1,850
HE shell_____do____	1,450

Muzzle energy_____foot tons__   349.2
Burning time of tracer_____seconds__   3
Maximum powder pressure__pounds per square inch__   36,000
Weight of bursting charge_____pounds__   1.47
Rifling:
    Length_____inches__   69.6
    Number of grooves_____   24
    Twist, uniform, R.H., one turn in_____calibers__   25.59
Weight of fixed round_____pounds__   19.36
Weight of projectile_____do____   14.4
Weight of powder charge_____do____   2.00
Travel of projectile in barrel_____inches__   72.54
Rate of fire_____rounds per minute__   20

b. General data pertaining to 75-mm tank gun M2 and mount.

Maximum elevation_____   19°12'
Maximum depression_____   7°48'
Turns of handwheel to elevate through maximum
    travel (27°)_____   24
Amount of traverse to left_____   14°0'
Amount of traverse to right_____   14°0'
Turns of handwheel to traverse through maximum
    travel (28°)_____   25½
One turn of traversing handwheel traverses to right
    or to left_____   1°6'
One turn of elevating handwheel elevates or de-
    presses_____   1°8'
Weight of elevating shield_____pounds__   442
Weight of horizontal rotor_____do____   875
Weight of sighting device installation_____do____   149
Weight of housing for mount (part of hull)___do____   1,900
Weight of gun, recoil mechanism and elevating shield
    at trunnions_____   1,636

■ 3. DESCRIPTION OF GROUPS.—*a. Gun group.*—(1) *Tube.*—
The tube (fig. 1) is formed of one piece of alloy steel. The
rear end of the bore is suitably tapered to form the chamber,
and from chamber to muzzle the bore is rifled with a uniform right hand twist of one turn per 25.59 calibers. The
exterior of the breech end is threaded to screw into the

breech ring, and the shoulder immediately forward of the threads contains a keyseat for the breech ring key. Forward from the breech ring the exterior is cylindrical and smoothly

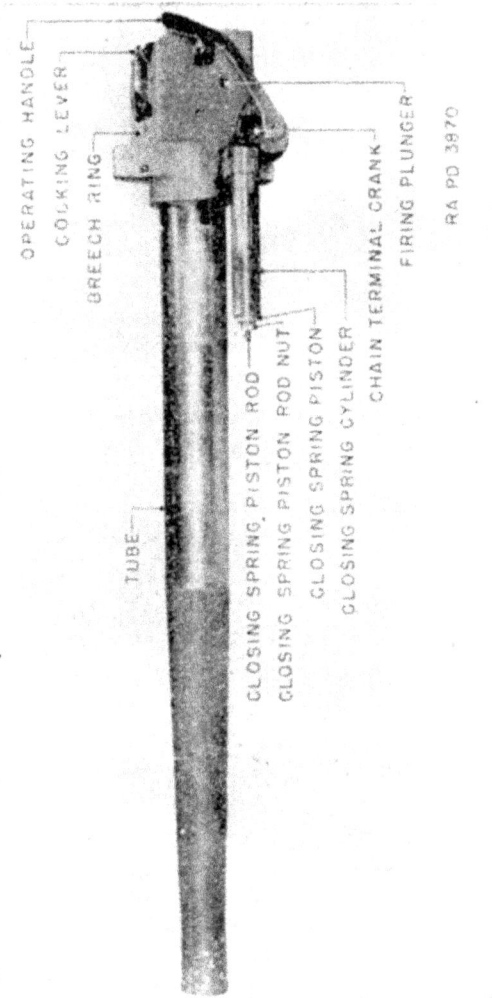

FIGURE 1.—75-mm tank gun M2—left side.

finished for bearing in the mount for approximately half the length of the tube. The breech face of the tube is recessed on each side of the bore to form extractor camming surfaces (fig 3).

(2) *Breech ring.*—(a) The exterior of the breech ring is rectangular in general form except the forward end which is

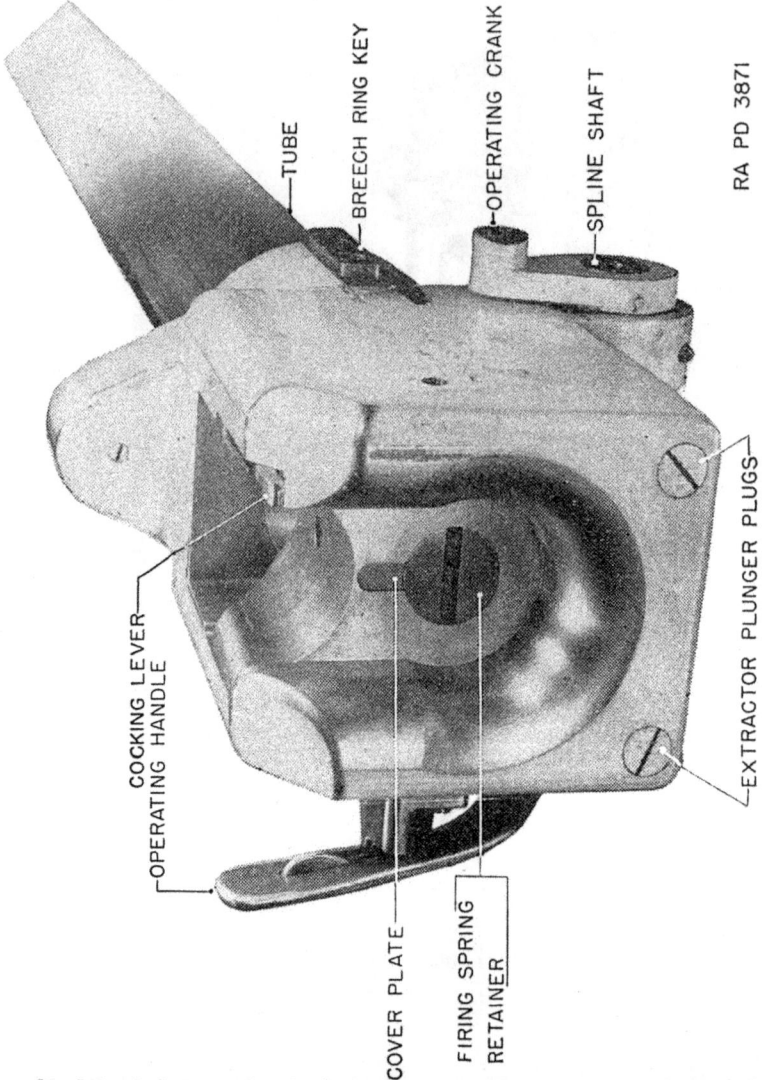

FIGURE 2 - 75-mm tank gun M2 - rear, breech in full closed position.

cylindrical for a short distance. Lugs projecting from the top and bottom near the front end are bored and threaded to receive the piston rods of the recoil and counterrecoil mechanism. The bottom lug extends rearward on either side of the bottom of the breech ring and is bored transversely to provide bearings for the operating and chain terminal cranks.

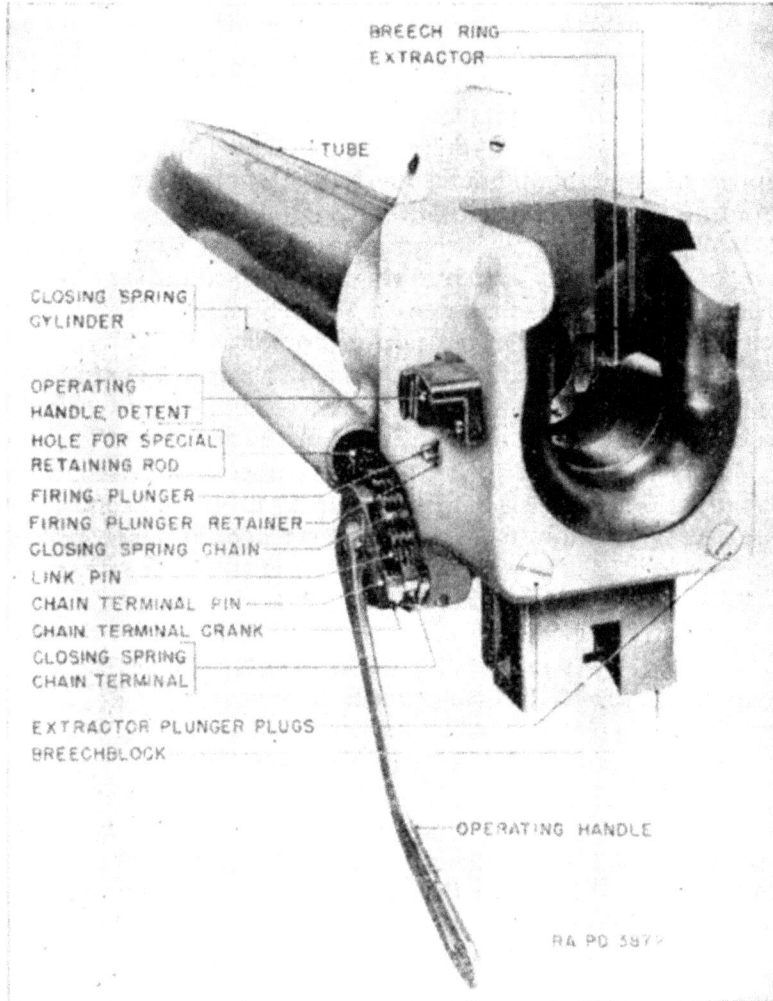

FIGURE 3.—75-mm tank gun M2—rear, breech in full open position.

An additional lug projects from the left side of the bottom lug and is bored longitudinally to receive the rear end of the closing spring cylinder. Tapped holes are provided for set

5

screws at piston rod and closing spring cylinder connections and drilled holes for tangential retaining pins at the crank bearing bores (figs. 2 and 3).

(*b*) The forward end of the breech ring is bored and threaded so that the tube may be screwed into it, and in the right face a shallow longitudinal keyseat has been cut for the breech ring key. At the front this keyseat terminates in a square notch through the front face of the breech ring, the notch alining with the keyseat in the tube. Two holes are tapped in the longitudinal keyseat so that the key may be held in place by two screws (fig. 2).

(*c*) The rectangular breech recess (fig. 3) is located in the rear portion of the breech ring and extends through from top to bottom. The rear wall of the recess has a U-shaped slot to clear the cartridge case in loading and ejecting. The right rear wall is reduced in height and sloped inward at the top to form a camming surface for the cocking lever. Inside the recess in the right and left walls are short curved recesses or slots which receive and guide the outer trunnions of the extractors. Two holes in the rear face of the breech ring, extending forward into the extractor trunnion recesses, house the extractor plungers and plunger springs. The rear ends of the holes are threaded for the extractor plunger plugs.

(*d*) A transverse hole is drilled, counterbored, and threaded through the lower left wall of the breech ring to receive the firing plunger and firing plunger retainer.

(3) *Breech ring key.*—The bronze breech ring key (fig. 2) has a rectangular body and a rectangular head on one end projecting to one side only. The key is held in place in the key seat of the breech ring by means of two cap screws, the head of the key extending inward through the notch in the front face of the breech ring and engaging the key seat of the tube. The body projects to the right of the breech ring and engages the keyway in the mount to prevent rotation between the gun and the mount.

(4) *Breech mechanism.*—The breech mechanism (fig. 4) is composed of the breechblock assembly, firing mechanism, extractors, spline shaft, breechblock crank, operating crank, closing mechanism, and related parts.

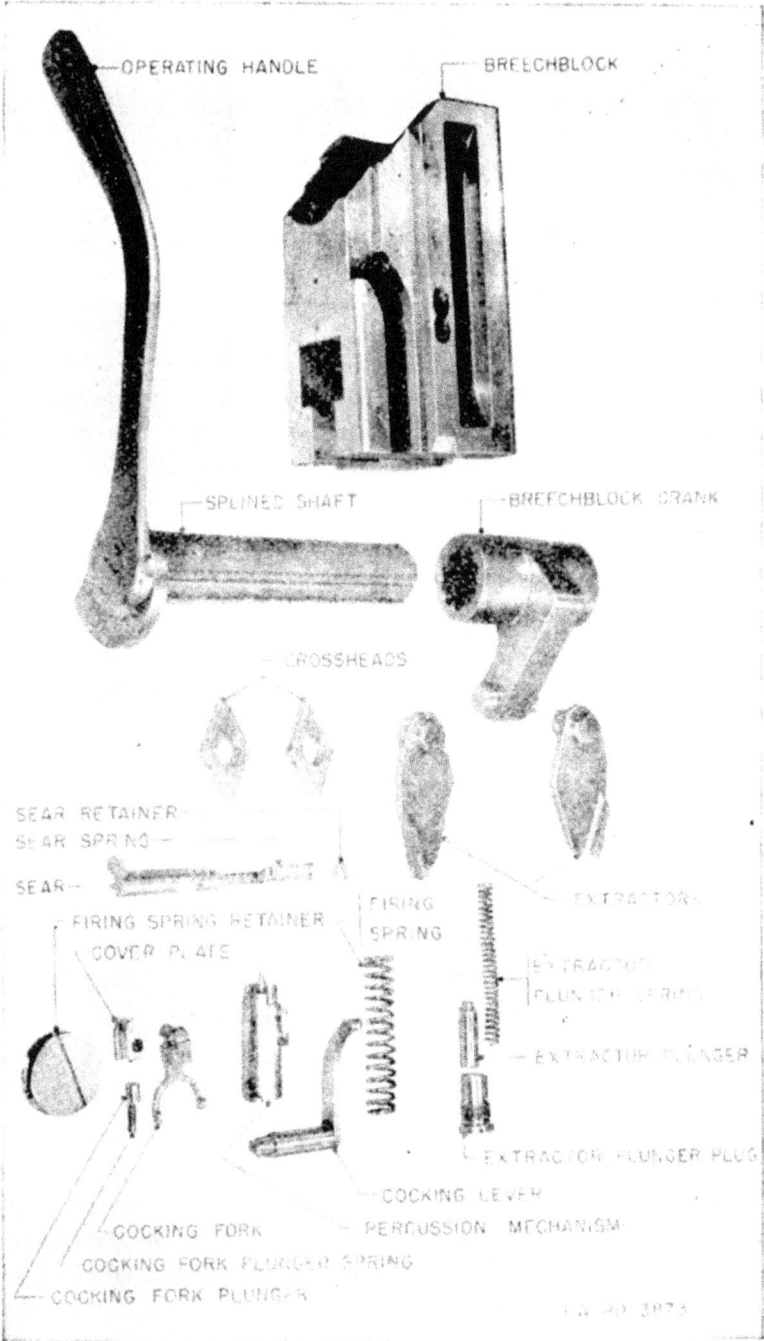

OPERATING HANDLE — BREECHBLOCK

SPLINED SHAFT — BREECHBLOCK CRANK

CROSSHEADS

SEAR RETAINER
SEAR SPRING
SEAR — EXTRACTORS

FIRING SPRING RETAINER — FIRING SPRING
COVER PLATE — EXTRACTOR PLUNGER SPRING

EXTRACTOR PLUNGER

EXTRACTOR PLUNGER PLUG

COCKING LEVER
COCKING FORK — PERCUSSION MECHANISM
COCKING FORK PLUNGER SPRING
COCKING FORK PLUNGER

FIGURE 4.—Breech mechanism unassembled.

7

(5) *Breechblock.*—(*a*) The breechblock (fig. 4) is of the vertical sliding type, having a shoulder on each side which slides inside a corresponding groove in each side of the interior of the breech ring. A semicircular channel has been hollowed out in the top of the breechblock to aline with the bottom of the U-slot in the rear wall of the breech ring and guide the cartridge into the chamber. To drive the cartridge into the chamber as the breechblock is raised, the upper front edge is beveled, and to complete the seating of the cartridge as the breech is closed, the rear face and guide shoulders of the breechblock and the corresponding surfaces of the grooves in the breech ring are inclined slightly so that the breechblock is cammed forward in rising. There is an inclined T-shaped slot in the lower end of the breechblock in which the lever of the operating shaft moves to raise and lower the breechblock. In each side of the breechblock is a groove or slot in which the inner trunnions of the extractors slide. The lower portion of the grooves is parallel to the breechblock guiding surfaces, but the upper ends are curved so that the extractors will be cammed fast enough to eject the cartridge case.

(*b*) A hole bored through the center of the breechblock houses the percussion mechanism. The forward end is counterbored and threaded to receive the breechblock bushing which is retained by a locking screw. The middle portion contains a longitudinal groove to receive and guide the sear lug on the firing pin guide. The sear is counterbored to receive the firing spring retainer, and on the upper side of this counterbore an inverted U-shaped recess houses the cocking fork and cover plate. A small vertical hole in the bottom of the counterbore receives a pin which, with the cover plate, locks the retainer in position. Arrows are engraved on the rear face of the breechblock to indicate the locked and unlocked positions of the retainer.

(*c*) A transverse bore through the breechblock, intersecting the lower side of the central bore, houses and guides the sear, sear spring, and sear retainer. An internal shoulder near the right end of the bore forms a seat for the sear spring and retainer, and a recess at the left end of the bore provides clearance for the arm of the sear. A shallow

V-shaped recess in the upper right side of the breechblock gives clearance for the arm of the cocking lever between the breechblock and breech ring, and a tranverse bore from this recess into the central cocking fork recess serves as a bearing for the shaft of the cocking lever.

(d) A groove is cut into each side of the breechblock to reduce weight, and a tapped hole in the top permits screwing in an eyebolt for disassembly and assembly.

(6) *Extractors.*—The extractors (fig. 4) are short levers supported vertically between the sides of the breechblock and the inner side walls of the breech recess by means of outer trunnions which bear in the short curved slots in the inner side walls of the breech recess and inner trunnions which ride in curved grooves in the sides of the breechblock. The extractors can thus rock against the front surface of the breech recess as the breechblock is raised or lowered.

(7) *Extractor plungers.*—The cylindrical extractor plungers (fig. 4) are pressed forward onto the outer trunnions of the extractors by means of the helical extractor plunger compression springs. Both plungers and springs are retained in the breech ring by threaded plugs which contain seats for the springs in their inner ends.

(8) *Spline shaft.*—The spline shaft (fig. 4) fits into the hubs of the operating crank, breechblock crank, and chain terminal crank, all three cranks rotating as a unit. There is a hole in the left end of the shaft for attachment of the operating handle by means of a screw. The right end of the shaft is drilled radially to provide a seat for the operating crank detent and to house the plunger. An axial hole provides access to the plunger for disengaging the detent, and a small tapped hole receives the plunger retaining screw.

(9) *Breechblock crank.*—The breechblock crank (fig. 4) is actuated and supported by the spline shaft which passes through it. The cylindrical pivot is located at the outer end of the crank arm and projects on either side to carry the bronze cross heads which slide in the T-slot of the breechblock. A hole from the upper and lower surfaces to the pivot hole provides for passage of lubricant. An arrow engraved on the side indicates the muzzle face.

9

(10) *Operating crank.*—The operating crank (fig. 2) consists of an internally splined hub with a short heavy arm on the right end. The splined interior of the hub receives the spline shaft while the cylindrical exterior is supported in bronze bushing in the transverse bore of the breech ring lug. The crank arm extends upward at the right side of the breech ring and terminates in an outward projecting lug in position for actuation by the operating cam on the mount. An annular groove is cut into the exterior of the hub so that the operating crank retainer pin can hold the crank in its bearing. The hub is drilled and counterbored radially to house the detent and spring which hold the shaft in lateral position.

(11) *Chain terminal crank.*—The chain terminal crank (fig. 3) consists of an internally splined hub with a short heavy crank arm on its left end. The end of the arm is slotted and drilled for attachment of the chain terminal. The splined interior of the hub receives the spline shaft while the cylindrical exterior is supported in a bronze bushing in the transverse bore of the breech ring lug. An annular groove is cut into the exterior of the hub for engagement by the pin which retains the crank in its bearing.

(12) *Closing spring mechanism.*—(a) The closing spring mechanism operates in the closing spring cylinder (fig. 1) which is supported by a lug on the left side of the breech ring immediately forward from the chain terminal crank. The closing spring piston rod, attached to the chain terminal crank by means of the closing spring chain, draws the piston rearward against the closing spring when the breech is opened.

(b) The closing spring cylinder (fig. 3) is a tube with a shoulder turned and drilled at the rear end to fit and lock into the supporting breech ring lug. An internal shoulder near the rear end forms a seat for the closing spring.

(c) The closing spring piston rod (fig. 1) is slotted and drilled at one end for attachment to the closing spring chain by means of a link pin. Near this end an additional hole drilled through the rod permits insertion of a tool in assembly and disassembly. A series of holes equally spaced is drilled along the threaded portion of the rod so that the piston rod nut can be prevented from unscrewing at various positions of adjustment by means of a cotter pin.

(d) The closing spring piston (fig. 1) is a bronze casting which fits into the closing spring cylinder bore freely. A hole centrally bored provides a loose fit for the piston rod. A concave recess at one end forms a seat for the piston rod nut while a shoulder at the other end provides a seat for the closing spring.

(e) The closing spring is a heavy helical compression spring.

(f) The closing spring piston rod nut (fig. 1) is a hexagonal nut with a special flange having a convex surface which fits into the concave recess of the piston to provide a flexible joint.

(g) The closing spring chain (fig. 3) is a commercial steel leaf or balance chain with a minimum strength of 5,500 pounds. The dimensions are as follows:

Height of side bars_____ 9/16 inch.
Pitch_____ 0. 620 inch.
Thickness of leaf_____ about 1 inch.
Width of chain_____. 0. 7 inch.

(h) The closing spring chain terminal (fig. 3) is a small steel tapered block with the thinner end slotted and drilled for attachment to the chain by means of a link pin, and the thicker end drilled for attachment to the chain terminal crank by means of the chain terminal pin.

(13) *Percussion mechanism.*—(a) The firing pin guide (fig. 5) is a cylindrical cup which slides axially in the center bore of the breechblock with its closed end forward. It carries the firing pin, retracting spring, firing spring stop, and the forward end of the firing spring. The larger of two exterior lugs on the lower side of the guide serves for engagement with the sear while the smaller lug nearer the front acts as a guide in the groove of the breechblock bore. One right hand and one left hand cocking lug extend outward from the guide at the rear end so that the cocking fork engages them and thus actuates the guide.

(b) The firing pin (fig. 5) is a shouldered screw with cylindrical body, slotted head, and flat point. It is screwed into the forward end of the guide and secured transversely by means of a headless pin.

(c) The firing spring stop (fig. 5), which has the form of a ring with two prongs protruding from its front face, fits

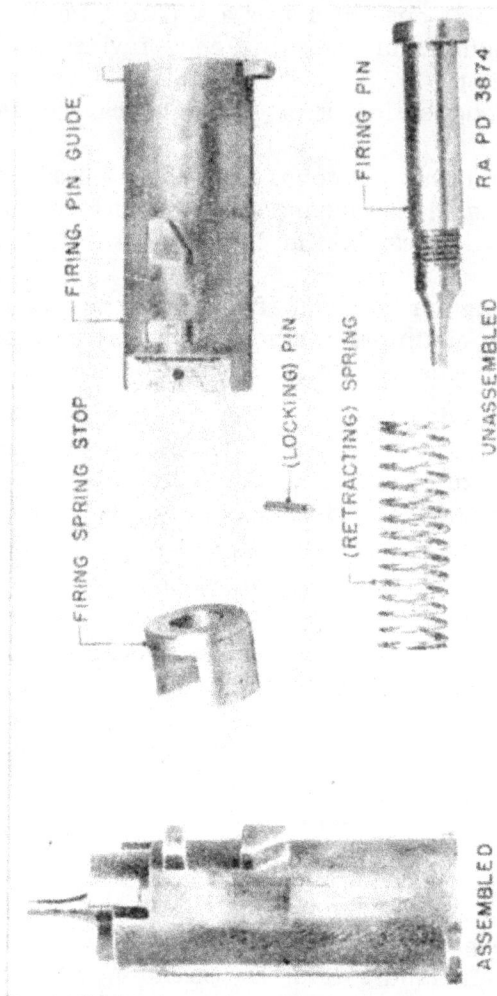

FIGURE 5.—Percussion mechanism.

freely in the annular space between the body of the firing pin and the interior surface of the guide with its prongs extending forward through openings in the end of the guide.

12

(*d*) The retracting spring (fig. 5) is a light helical compression spring mounted on the body of the firing pin. It bears rearward on the firing pin head and forward on the firing spring stop, pressing the stop against the forward end of the guide.

(14) *Firing spring.*—The firing spring (fig. 4) is a helical compression spring which extends into the firing pin guide from the rear in the annular space between the retracting spring and the interior surface of the guide. Its forward end bears on the firing spring stop, and the rear end seats in the recess of the firing spring retainer.

(15) *Firing spring retainer.*—The firing spring retainer (fig. 4) is a cylindrical plug which closes the rear end of the central bore in the breechblock. Its inner end, besides being recessed axially to form a seat for the firing spring, carries two segmental lugs which engage a pin in the breechblock and a lug on the cover plate to lock the retainer in place. The rear face of the retainer is slotted to facilitate turning with fingers and is marked with an arrow on the bottom vertical center line as an aid in assembly and disassembly.

(16) *Sear.*—The sear (fig. 4) is a cylindrical bar which slides transversely in the breechblock across the path of the sear lug of the firing pin guide. It is notched in its upper side for engagement and release of the sear lug of the firing pin guide, and reduced in diameter on the right end to receive the helical compression sear spring which seats in the right side of the breechblock and presses the sear to the left. The right end of the sear is grooved circumferentially to receive the sear retainer. The left end of the sear is formed into a downward projecting arm which bears in a recess in the breechblock to prevent rotation and provides a contact surface for the firing plunger.

(17) *Cocking fork.*—The cocking fork (fig. 4) consists of a hub with forked arm which straddles the firing pin guide to contact the cocking lugs on either side of the firing pin guide and force it rearward to cocked position. The shaft of the cocking lever passes through the hub of the cocking fork.

(18) *Cocking lever.*—The cocking lever (fig. 4) consists of a cylindrical shaft shouldered and flattened at one end to

fit the hub of the cocking fork, and a curved arm at the other end. A transverse bore in the breechblock contains the shaft, and the curved arm of the cocking lever extends upward and rearward between the right side of the breech ring, projecting over the rear wall of the breech ring.

(19) *Cover plate.*—The cover plate (fig. 4) is machined to fit and close the rear opening of the breechblock recess above the firing spring retainer. A convex curve at the top fits the recess, a concave circular bottom edge rests on the retainer, and flanges at either side engage the inner rear wall of the recess in the breechblock. A lug projecting downward on the bottom retains the upper side of the firing spring retainer. The forward or inner face of the plate is drilled to house the cocking fork plunger and spring.

(20) *Cocking fork plunger.*—The cocking fork plunger (fig. 4) is cylindrical in form and bears forward on the lower side of the cocking fork hub to return the cocking fork and cocking lever to starting position as the cocking lever is released. The rear end of the plunger is recessed to receive the front end of the helical compression plunger spring which presses forward on the plunger and rearward on the cover plate.

(21) *Firing plunger.*—The firing plunger (fig. 3) is cylindrical in form with an integral collar near its middle. It is contained in a hole bored through the lower left side of the breech ring. Its flat inner end in alined with the contact surface on the arm of the sear when the breechblock is in closed position, and its rounded outer end projects from the breech ring in position for operation by the firing mechanism on the mount. The plunger is retained in the breech ring by the firing plunger retainer which screws into the breech ring around the plunger and confines the collar.

b. Mount group.—(1) *Cradle.*—The cradle (fig. 8) rocks on its trunnions on roller bearings which are supported in the trunnion seats of the horizontal rotor. The bore of the cradle is provided with a bronze liner through which the gun slides in recoil and counterrecoil. The recoil cylinders are mounted above and below the bore of the cradle while an extension on the lower right rear end contains the operating crank camming surface which automatically opens the breech. A support on the left side of the cradle (fig. 12) holds the solenoid, firing lever link, and firing lever which comprise the firing mechanism on the mount. Two arms on the left side and one on the right project from the sides of the cradle. Holes bored through them provide a means for bolting the shoulder guard (not shown) to the cradle.

(2) *Recoil and counterrecoil system.*—(*a*) *General.*—The function of the recoil system (fig. 8) is to check the movement of the recoiling mass in a gradual manner so as not to displace the mount and to return the gun to battery without shock at all angles of elevation in order that the gun may be fired again.

(*b*) The recoil piston rods are attached to the breech ring and recoil with the gun when the piece is fired.

(*c*) All space in the recoil cylinders not otherwise occupied is filled with heavy recoil oil, and this oil in the rear of the recoil pistons must pass through grooves in the cylinder during recoil in order to pass from the rear to the front of the piston.

(3) *Traversing mechanism.*—The traversing mechanism (fig. 10) consists of a traversing handwheel gear case and handwheel connected to the traversing pinion shaft housing and pinion by means of a shaft which transmits the motion of the handwheel through the gears to the pinion. The handwheel gear case is bolted to the horizontal rotor (fig. 11) on the left side at the bottom. The pinion shaft housing is bolted to the rotor on the right side at the bottom. Turning the traversing handwheel causes the pinion to mesh with the traversing rack affixed to the tank and rotate the horizontal rotor about its vertical axis. The firing button, located at the center of the handwheel, actuates the switch lever affixed to the bottom of the handwheel gear case.

(4) *Elevating mechanism.*—The elevating mechanism (fig. 9) is mounted on the horizontal rotor (fig. 11). When the elevating handwheel is turned, the motion is transmitted through a train of bevel gears to the pinion which meshes with the elevating rack bolted to the elevating shield.

(5) *Horizontal rotor.*—The horizontal rotor (fig. 6) is that part of the mount which rotates about its vertical axis when the gun is traversed. The gun, elevating shield, and recoil mechanism are supported by the trunnion in the trunnion

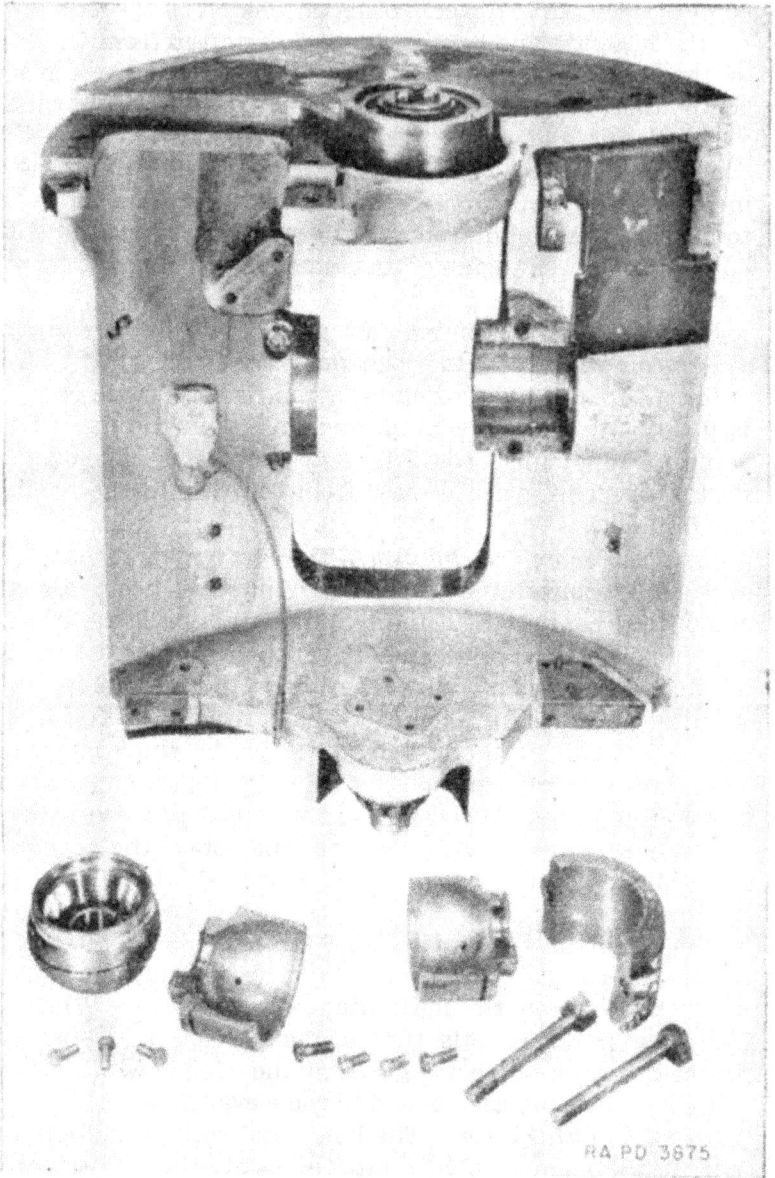

FIGURE 6.—Horizontal rotor—rear view.

seats of the rotor, and the whole assembly is supported by a vertical thrust roller bearing on the bottom of the rotor. Another roller bearing at the top supports the rotor vertically. A rectangular opening permits the elevating shield to extend

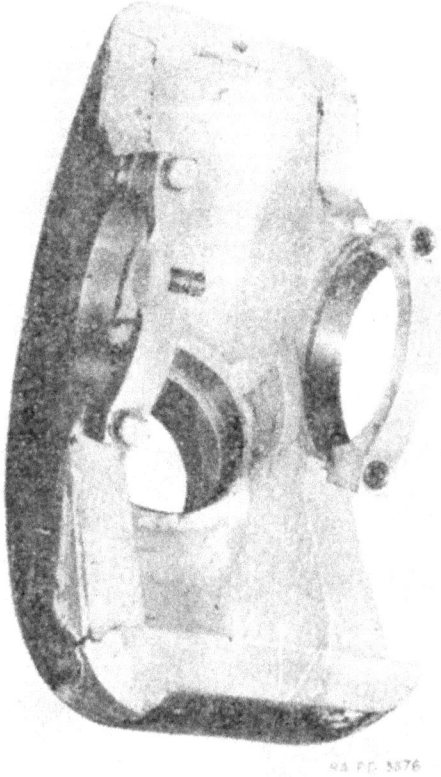

FIGURE 7.—Elevating shield—rear view.

through the rotor and rotate about the horizontal axis of the trunnion seats.

(6) *Elevating shield.*—The elevating shield (fig. 7) is that part of the mount which retains and protects the recoil mechanism. It is connected to the recoil mechanism trunnions, the gun tube passing through a smooth bore in the

17

TRUNNION BEARING BUFFER COVER

RECOIL CYLINDER (TOP)
SOLENOID
PISTON ROD FOLLOWER
LUG NUT WASHER
LUG NUT
PISTON ROD NUT

CYLINDER COVER
FIRING LEVER LINK

FIRING LEVER

OPERATING CRANK CAM

RA PD 3877

CRADLE

RECOIL CYLINDER (BOTTOM)

FIGURE 8 - Recoil mechanism - left side

18

center of the shield. Almost the entire cylindrical surface of the shield projects beyond the horizontal rotor through a rectangular opening in the rotor.

FIGURE 9.—Elevating mechanism—front view.

(7) *Shoulder guard.*—The shoulder guard is for the protection of the gunner. It shields the firing mechanism on the mount and extends about 14 inches to the rear of the rear face of the breech.

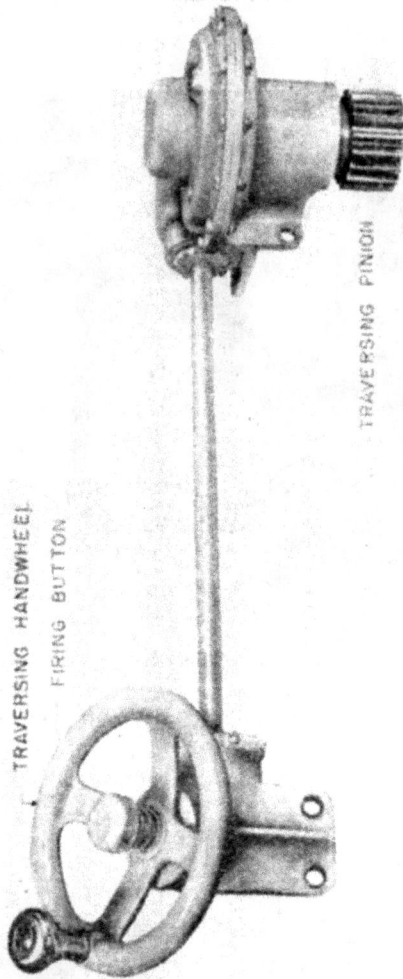

FIGURE 10.—Traversing mechanism.

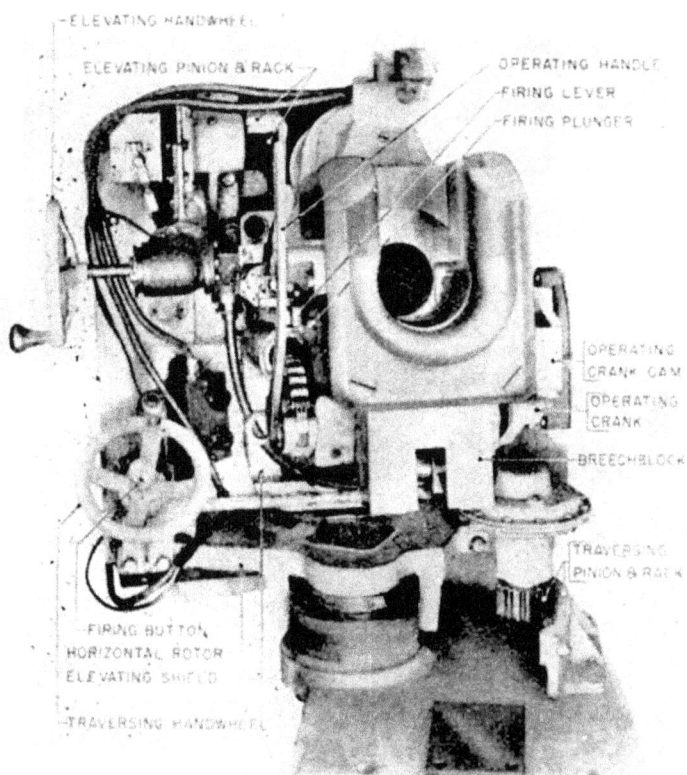

FIGURE 11.—75-mm tank gun M2 and mount in medium tank M3—
rear view.

FIGURE 12.—Firing mechanism on mount in medium tank M3—
left side.

Section II

DISASSEMBLY AND ASSEMBLY

■ 4. GENERAL.—Disassembling may be considered under two general heads: removal of parts to the extent required for ordinary cleaning and repairs in the field; and detail disassembling involving the removal of additional parts when it is necessary to perform more detailed cleaning or to adjust or replace parts.

■ 5. FIELD DISASSEMBLY.—*a. Shoulder guard.*—With a 1¼ open-end wrench, loosen the three nuts that secure the shoulder guard to the cradle and remove the nuts and washers. Remove the guard from the cradle by pulling it to the rear. Place the washers and nuts on the guard and place the guard in a convenient place.

b. Extractor plungers.—Unscrew the extractor plunger plugs from the lower rear face of the breech ring, using care to prevent the plunger springs from throwing the plugs to the rear. Remove the plugs, springs, and plungers. (These parts are interchangeable.)

c. Firing spring retainer.—Actuate the trigger to be sure the firing spring is not compressed. Press the firing spring retainer into the breechblock as far as it will go. Rotate it counterclockwise until the arrow on the face of the retainer is alined with the arrow marked "Open" on the rear face of the breechblock. Release the pressure and the retainer will be forced out by the firing spring.

d. Firing spring and cover plate.—Support the cover plate with the index finger and withdraw the firing spring. Slide the cover plate downward from the breechblock using care not to drop the cocking fork plunger and spring from the plate.

e. Firing pin guide assembly.—Depress the firing plunger so that the sear notch will be in line with the sear guide lugs on the firing pin guide. Cup one hand behind the opening in the rear of the breechblock and rotate the cocking lever forward smartly. The firing pin guide assembly will be moved far enough to the rear by the cocking fork to enable removing it with the fingers.

f. Breechblock.—(1) Insert a ¼-inch rod into the hole in the right end of the spline shaft. Pull the rod to the rear and move the detent release plunger far enough to the rear to disengage the spline shaft detent, at the same time drift the shaft to the left far enough to clear the detent. Remove the rod from the shaft.

(2) Engage the left end of the spline shaft with the operating handle or wrench and lower the breechblock to the point where the ¼-inch rod may be placed between the shoulder of the chain terminal crank and the rear end of the closing spring cylinder lug. Hold the rod in place while releasing the pressure on the closing spring. This releases the tension exerted on the spline shaft by the spring.

(3) Removing the breechblock from the breech ring is a two-man job—one man removes the spline shaft then supports the extractors, and the other man supports and removes the breechblock and breechblock crank. As one man pulls the shaft to the left the other man uses both hands underneath the breech ring to support the breechblock and crank. After the shaft has been removed the hub of the breechblock crank is removed from its recess in the lower breech ring lug, and the breechblock and crank are eased out through the bottom of the breech ring. As the breechblock is being removed the extractors are supported and are removed as soon as possible.

g. Extractors.—To remove the extractors, grasp them with the fingers, rotate them to a vertical position, and pull the outer trunnions from the curved slots in the breech ring.

h. Breechblock crank.—Slide the bronze crossheads out of the T-slot in the bottom of the breechblock to remove the crossheads from their pivots and remove the breechblock crank.

i. Cocking lever and fork.—Support the cocking fork from inside the firing pin guide assembly recess, at the same time pull the cocking lever to the right. Remove the cocking fork.

j. Sear.—Place the breechblock front face down. With the index finger press the sear into the block far enough to expose the sear retainer on the opposite side. Slide the sear retainer from the right end of the sear and release the pressure. Remove the sear and sear spring.

■ 6. FIELD ASSEMBLY.—*a. Sear.*—Place the breechblock front face down. Slide the sear spring over the small end of the sear and insert the sear and spring into the sear recess in the left side of the breechblock. Press the sear into the block far enough to permit the engagement of the sear retainer on the right end of the sear. Be sure the retainer seats in the counterbore when the pressure on the sear spring is released.

b. Cocking lever and fork.—Insert the cocking fork into its recess, above the firing pin guide assembly recess, with the hub upward, slot to the left, and the rounded lugs of the fork to the rear. Hold the fork in a vertical position and insert the shaft of the cocking lever into its recess in the right side of the breechblock, the arm of the lever up. Push inward until the end of the shaft enters the hub of the cocking fork. Swing the lever forward and rearward through a small angle while pushing to the left until the shaft fully seats in the hub of the fork.

c. Breechblock crank.—Place the bronze crossheads on their pivots on the crank and insert them into the front end of the T-slot in the breechblock, offset hub of the crank forward. Be sure the crossheads are facing properly with the index arrow toward the muzzle and the openings of the oil holes against the bearing surfaces of the T-slot.

d. Extractors.—Place the extractors in position in the breech recess with their long outer trunnions in the short curved slots in the sides of the breech ring and their lips in the pockets on either side of the chamber.

e. Breechblock.—(1) Support the breechblock and crank by holding the breechblock at the bottom and the crank between the fingers. Rotate the cocking lever forward and start the breechblock up into the breech recess. Raise the block until it is stopped by the extractors which are being held in the breech ring by another man. Rotate the lips of the extractors forward and raise the breechblock to the closed position.

(2) Swing the hub of the crank forward into its recess in the lower breech ring lug and aline the splines of the breechblock crank with those of the chain terminal crank. Start the spline shaft into the chain terminal crank, with the detent release plunger to the right, far enough to support the breech-

block. Aline the splines of the operating crank and continue moving the shaft to the right until the detent seats in its recess in the spline shaft.

(3) With the operating handle or wrench, turn the chain terminal crank to the rear and remove the ¼-inch rod from between the crank and the closing spring cylinder lug. Gradually release the pressure on the handle or wrench and allow the breechblock to rise to the closed position.

f. Firing pin guide assembly.—Insert the firing pin guide assembly into its recess in the rear face of the breechblock, striker end of the firing pin first with the guide lugs on the bottom of the firing pin guide in alinement with the groove in the bottom of the recess. Depress the firing plunger to bring the sear notch in line with the groove and push the guide assembly forward until it strikes the breechblock bushing. Release the pressure on the firing plunger.

g. Cover plate and firing spring.—If the cocking fork plunger and spring have been removed from the cover plate, place the spring inside the cocking fork plunger and insert both, spring end first, into the hole in the cover plate. Insert the plate into its recess just behind the hub of the cocking fork with the plunger to the front and the concave surface down. Hold the plate in position with the index finger and insert the firing spring into the firing pin guide.

h. Firing spring retainer.—Place the cupped end of the firing spring retainer over the rear end of the firing spring. Aline the arrow on the rear face of the retainer with the arrow on the rear face of the breechblock marked "Open" and push it forward as far as possible. Rotate the retainer clockwise until the arrow on it is alined with arrow marked "Lock" and then release the pressure.

i. Extractor plungers.—Place the small ends of the extractor plunger springs over the small ends of the extractor plungers and insert both plungers and springs into their holes in the lower rear face of the breech ring with the plungers forwards. Retain the springs by screwing in the extractor plunger plugs.

j. Shoulder guard.—Engage the studs of the guard with their holes in the cradle. Replace the washers and nuts and tighten with a 1¼-inch wrench.

■ 7. DETAIL DISASSEMBLY.—Detail disassembly consists of all the operations listed in paragraph 5 with the addition of operations in *a* to *c*, inclusive, below. Detail disassembly may be performed by using services in whole or in part as the situation demands but should not be included as part of the daily training exercises.

a. To disassemble the firing pin guide assembly, press the prongs of the firing spring stop into the guide and drift out the exposed guide pin. Release the stop and unscrew the firing pin from the guide. Remove the firing pin, retracting spring, and firing spring stop.

b. To remove the chain terminal crank, insert the spline shaft into the hub of the chain terminal crank. With the operating handle or wrench rotate the crank to the rear until the hole in the rear end of the closing spring piston rod emerges from the closing spring cylinder. Insert a ¼-inch drift through the piston rod so both ends of the drift rest against the rear end of the closing spring cylinder lug. Gradually release the pressure on the operating wrench leaving the closing spring compressed. Remove the spline shaft. Remove the cotter key from the chain terminal crank and pin. Drift the chain terminal pin to the right and disengage the chain terminal from the crank. Unscrew the crank retaining pin from the crank bearing and pull the crank to the left from the breech ring.

c. To remove the operating crank, level the gun with the elevating handwheel and *keep it level all the time the piston rods are disengaged* from the breech ring. This is a safety precaution. With an Allen setscrew wrench loosen the Allen setscrews that lock the piston rod nuts by giving each nut several turns at a time. After the piston rods are loosened, slide the barrel assembly about 6 inches to the rear (far enough to clear the operating cam). Remove the operating crank retaining pin with a screw driver and pull the crank from its bearing.

d. To replace the operating crank, make sure the hub and bearing are clean, then start the hub of the operating crank into its bearing on the right side of the lower breech ring lug. Move the crank to the left until it is fully seated against the breech ring then replace the crank retaining pin. Ease the barrel assembly forward while keeping the piston rod nut recesses alined with the piston rod nuts. *Do not, under any circumstance, allow the barrel assembly to go forward with a slam. Ease it forward.* Seat the piston rod nuts in their recesses and tighten them. Tighten the Allen setscrews.

e. To replace the chain terminal crank, place the hub of the chain terminal crank into its bearing, move the crank up against the breech ring, and replace the retaining pin. Hold the chain terminal in its recess in the top of the crank arm and insert the chain terminal pin from the right, through the terminal and crank. Aline the holes in the pin and the crank arm and insert the cotter key from the top. Spread the end of the key. Insert the spline shaft into the hub of the crank, remove the ¼-inch rod from the closing spring piston rod, place it between the chain terminal crank and the closing spring cylinder lug, and gradually release the pressure on the closing spring.

f. To assemble the firing pin guide assembly, place the firing spring stop into the guide with the prongs toward the closed end so they will protrude through the half-round holes. Insert the firing pin into the retracting spring and place both the firing pin and spring into the guide with the threads of the firing pin to the front.. Screw the firing pin into the guide as far as it will go then back it out until the holes of the pin and guide are alined. Push the prongs of the firing spring stop into the guide and replace the guide pin. Release the pressure on the firing spring stop.

SECTION III

MECHANICAL FUNCTIONING

■ 8. GENERAL.—The subject of mechanical functioning is divided into different phases to make description easier to present and for the student to understand. However, it must be remembered that most of these phases take place at the same time. The first phase starts just as the gun returns to battery.

■ 9. FIRST PHASE—AUTOMATIC OPENING OF BREECH.—*a.* Just before the gun reaches the full forward position, the projecting lug on the arm of the operating crank strikes the operating cam and rotates the operating crank to the rear. This motion is transmitted through the spline shaft to the breechblock and chain terminal cranks. As the arm of the breechblock crank moves to the rear, the cross heads move downward to the rear in the inclined T-slot in the bottom of the breechblock, and bring the block to its lowermost position. This downward motion is stopped by the stop surface on the breechblock crank striking the breech ring.

b. During the downward movement of the breechblock the chain terminal crank is rotated to the rear compressing the closing spring between the closing spring piston and the rear end of the closing spring cylinder. After the breechblock reaches its lowermost position, it starts to move up under the action of the closing spring. This movement is stopped and the breechblock is locked in the open position by inner trunnions of the extractors engaging the flat surfaces of the extractor cam slot shoulders.

■ 10. SECOND PHASE—COCKING ACTION.—As the breechblock moves down in its recess the upper arm of the cocking lever is cammed forward by the cam surface inside the breech ring. This motion is transmitted to the cocking fork through the cocking lever shaft and moves the fork to the rear, compressing the cocking fork plunger and spring. The cocking fork engages the cocking lugs on either side of the firing pin guide, forces the guide to the rear, and thus compresses the firing spring between the firing spring stop and the firing spring retainer. The firing pin guide is moved to the rear far enough to permit the sear, under the action of the sear spring, to move in front of the sear lug and hold the guide in its rearmost position.

■ 11. THIRD PHASE—EXTRACTION.—As the breechblock nears its lowermost position the inner trunnions of the extractors, riding in the extractor cam slots in each side of the block, are cammed forward. The extractors are pivoted about their elbows and impart a sharp rearward throw to their lips. The lips of the extractors being in front of the rim of the shell, the empty case is extracted and ejected clear of the breech ring. During this movement the inner trunnions of the extractors move out of the extractor cam slots, seat themselves on the flat surfaces on the extractor cam slot shoulders, and hold the breechblock open. The extractor plungers and springs, acting on the outer trunnions, hold the lower portion of the extractors forward to insure the engagement of the inner trunnions and the extractor cam slot shoulders.

■ 12. FOURTH PHASE—CLOSING THE BREECH.—*a.* As a live round is shoved into the chamber, the rim strikes the lips of the extractors rotating the upper portion of the extractors forward and compressing the extractor plunger springs. This movement disengages the inner trunnions of the extractors from the extractor cam slot shoulders permitting the breechblock to rise under the action of the closing spring.

b. As the breechblock rises the bevel on the front face forces the live round into the chamber. The cocking lever and fork, under the action of the cocking fork plunger and spring, return to the normal position with the cocking lever protruding from the breech ring and the cocking fork forward clear of the path of the cocking lugs. The firing pin guide assembly is retained in the cocked position by the sear. The upward movement of the breechblock is stopped by the breechblock crank arm striking the stop surface of the breech ring. The breech is closed and the gun is ready to fire.

■ 13. FIFTH PHASE—FIRING THE PIECE.—*a.* When the firing button in the center of the traversing handwheel is pressed, a switch is closed and an electrical circuit is completed. This creates an electromagnetic force which moves the solenoid plunger to the rear. The solenoid plunger contacts the firing lever link which in turn acts on the firing lever to cam the firing plunger into the breech ring. When the manual trigger button, which is forward and above the traversing handwheel, is used, a flexible cable running from the button to the solenoid plunger also moves the plunger to the rear. The firing lever link and firing lever are actuated and the firing plunger is forced into the breech ring.

b. As the firing plunger is cammed into the breech ring, it contacts the sear arm and moves the sear to the right against the action of the sear spring. As the sear moves to the right the sear notch comes in line with sear lug of the firing pin guide and thus permits the firing pin guide assembly to move forward under the action of the compressed firing spring to fire the round.

■ 14. SIXTH PHASE—RETRACTION OF THE FIRING PIN.—The compressed firing spring, bearing on the pronged stop which is seated in the forward end of the firing pin guide, forces the stop, guide, and firing pin forward as a unit. Just before the firing pin strikes the primer, the forward projecting prongs of the stop strike the rear of the breechblock bushing, stopping the action of the firing spring. The momentum of the firing pin and guide carries the firing pin forward to strike the primer, this final motion drives the stop into the guide to compress the firing pin retracting spring between the stop and the head of the firing pin. The retracting spring then expands to push the guide and firing pin rearward, thus withdrawing the firing pin from the chamber so the point will be flush with or slightly in rear of the breechblock bushing.

■ 15. SEVENTH PHASE—BACKWARD MOVEMENT OF THE BARREL ASSEMBLY (RECOIL).—a. As the gun is fired, the expanding gases drive the barrel assembly to the rear about 11½ inches. As the piston rods of the two recoil cylinders (top and bottom) are attached to the breech ring, they are pulled to the rear. During this movement the four counterrecoil springs are compressed between the piston brackets and the rear end of the cylinders as the pistons are forced from front to rear through the recoil oil.

b. As the pistons move to the rear the oil has only one way to get to the front, that is around the piston heads through the half-round tapered grooves in the bottom and top of the cylinders. The action of the oil and counterrecoil springs bring the gun to rest momentarily in full recoil.

c. The shoulder on each piston rod just behind the piston bracket enters a reservoir in the rear of the recoil cylinder forming an oil lock and thus preventing a metal to metal contact should the gun recoil more than 12 inches. This is a safety feature of the recoil mechanism.

■ 16. EIGHTH PHASE—FORWARD MOVEMENT OF THE BARREL AS-SEMBLY (COUNTERRECOIL).—The gun stops momentarily when it reaches the full length of recoil then is immediately forced forward by the compressed counterrecoil springs. The pistons move from rear to front and the oil returns to the rear of the piston heads by means of the two half-round grooves in each cylinder. Just before the gun reaches the full forward position, the tapered counterrecoil buffers in the front ends of the recoil cylinders enter the holes in the center of the piston rods. The oil is forced back into the cylinders through the decreasing ring shaped space between the tapered buffers and the recesses of the piston rods. The restriction of the oil going around the piston heads through the tapered half-round grooves in the two cylinders, and the added last movement restriction by the tapered buffers, bring the gun to rest in battery without appreciable shock.

SECTION IV

INSPECTIONS, ADJUSTMENTS, LUBRICATION, CARE AND CLEANING

■ 17. GENERAL.—Cannon become fouled with firing; moisture, dirt, and grease collect on the outside of the gun and in the moving parts. In order to keep the gun in efficient operating condition it is essential that the gun be cleaned properly, lubricated, and kept in adjustment. Only by periodic de-tailed inspections can a commander assure himself that these essential functions are being properly performed and corrections made before serious trouble ensues. By proper preventive maintenance the gun and its component parts will not only have a longer life but will be ready to function at all times.

■ 18. INSPECTIONS.—a. The company commander makes pe-riodic inspections, at least once per week in garrison and oftener in the field, to note the general appearance of the guns and to see that they are being properly maintained. The platoon leader assisted by the tank commander and the gunner make the detailed inspection of the gun at least once per week and oftener when the gun is being fired regu-larly. This inspection consists of the following:

Parts to be inspected in order of inspection	Points to observe
(1) Gun as a unit.	(1) Note general appearance and smoothness of operation of breech mechanism. Disassemble breech mechanism and firing mechanism. Inspect parts for wear, burs, or other mutilations. See that they are clean and well lubricated. Note condition of bore for copper deposits on the lands and in the grooves; erosion at origin of rifling. Examine bearing surface of exterior of tube for scoring or other mutilation. Check breech ring key for wear or mutilation and its screws for tightness.
(2) Breech recess.	(2) Note where there are scores or burs on bearing surfaces.
(3) Breechblock.	(3) Note whether there are scores or bruises on bearing surfaces, and whether breechblock bushing screw is secure and flush with or below front face of breechblock.
(4) Mount.	(4) Note general appearance and color and check all moving joints and bearings for lubrication.
(5) Recoil mechanism.	(5) Check the proper amount of recoil oil and inspect recoil cylinder for oil leaks. Checks all external nuts and fittings.
(6) Elevating mechanism.	(6) Elevate and depress the gun through the full extent of its travel. Note whether the mechanism operates without binding or excessive backlash.
(7) Traversing mechanism.	(7) Traverse the gun and mount throughout its complete travel. Note whether operation is without binding or excessive backlash.

b. When the gun is in daily use the gunner makes a daily inspection for cleanliness of gun and mount and the subjects covered in *a*(5), (6), and (7) above.

■ 19. ADJUSTMENTS.—*a. Closing spring.*—Approximate adjustment of the closing spring is obtained by tightening the closing spring piston rod nut until five cotter pin holes are exposed. Accurate adjustment is then made by trial. Open the breech. Trip the extractors. If the breechblock does not fully close or action is sluggish, tighten the nut one cotter pin hole at a time until the proper adjustment is reached. If the breechblock closes too fast loosen the nut one cotter pin hole.

b. Oil.—For proper amount of oil in the recoil cylinders see paragraph 21.

■ 20. RECEIVING NEW GUN.—*a.* When a new tank is received by the using service the rust-preventive compound that has been applied to the 75-mm gun should be removed immediately and the gun checked. The gun should be detail disassembled and all the parts cleaned with dry-cleaning solvent. The parts of the breech mechanism should be soaked in the solvent while the bore and breech ring are being cleaned. To clean the bore, run dry rags or waste through it in order to remove most of the rust-preventive compound before using the dry-cleaning solvent. Soak a rag in the solvent and run it through the bore until all of the rust-preventive compound has been removed. The chamber and breech ring should be cleaned in the same manner. The barrel bearing, elevating and traversing mechanisms, and all other exposed metal parts should be cleaned. The rust-preventive compound must be rinsed away with water after all the parts of the gun are free. Hot water is preferable. Dry all the parts thoroughly and then coat with oil.

b. During the cleaning of the gun inspect each part. Should any parts be missing or not in proper condition, check to see if this fact is mentioned in the gun book. If there is no mention of this condition, the ordnance should be notified immediately. Any missing or damaged parts are replaced and the gun assembled.

c. The operation of the elevating, traversing, and sight adjusting mechanisms should be checked as well as the functioning of the breech and firing mechanisms.

d. Usually when a new gun is received the recoil cylinders will be filled, but an inspection must be made to verify the amount of recoil oil in the cylinders.

■ 21. FILLING AND REPLENISHING RECOIL CYLINDERS.—a. When the recoil cylinders are filled with recoil oil (R1XS–121, recoil oil, heavy) for the first time or after draining, the following operations should be performed:

(1) Level the gun and remove the top rear filler plugs from both recoil cylinders.

(2) Fill the cylinders (by using a funnel or a recoil oil gun) until the oil reaches the top of the filler plug holes.

(3) Tap the cylinders with a block of wood to drive out any air that might be in the cylinders.

(4) Depress the muzzle as far as possible (9°) and allow the oil to run out of the cylinders. This establishes a void which will allow for the expansion of the oil when it becomes hot during firing.

(5) Replace the filler plugs and wipe the excess oil from the cylinders.

b. Oil is replenished by depressing the muzzle (9°), removing the top rear filler plugs, and adding oil until the oil level reaches the bottom of the filler plug holes. Replace and tighten the filler plugs.

■ 22. CARE DURING NORMAL TRAINING PERIOD.—a. When tanks arc used in the field exercises, sufficient time must be allowed for the cleaning of the tank guns. During training exercises the guns pick up dust or moisture, depending on the weather. After each such exercise, the gun must be field disassembled, cleaned with hot soapy water, rinsed, dried, and all exposed metal covered with oil. A rough rag should be used in cleaning the bore; burlap is one of the best materials. Any machine oil of medium grade will protect the exposed metal. During hot weather SAE No. 20, and during cold weather SAE No. 10 is desirable. A light cup grease will also serve the purpose during hot weather.

b. During damp or rainy weather the gun must be inspected daily. The oil must be removed from the bore to make a proper inspection. If rust or corrosion are found, the gun must be disassembled and thoroughly cleaned. Rapid changes in weather conditions may cause the gun to sweat and rust under the coating of oil. During these periods of changing weather, daily inspection should be made.

c. Each gun must be field disassembled at least once a week, whether it has been used or not, and thoroughly cleaned. One afternoon or morning should be set aside each week for this purpose during the training period.

d. Before a tank is deadlined for repairs, the gun crew should thoroughly clean the gun and apply a heavy coat of oil to all the exposed metal parts. During the summer a light cup grease will serve the purpose.

e. During the training period the responsibility of caring for the gun should be placed on the tank commander and gun crews. Frequent unexpected inspections should be held by the company officers.

■ 23. Care Before, During, and After Firing (training period).—*a.* Before firing the 75-mm tank gun, the following points should be checked:

(1) Check the bore and chamber to make sure they are clean and clear.

(2) Check the functioning of the breech mechanism by opening and closing the breech. Check the side movement of the operating cam.

(3) Check the operation of both the electric and manual trigger mechanisms.

(4) Check the operation of the traversing and elevating mechanisms.

(5) Make sure the piston rod nuts are properly engaged to the breech ring and that the recoil cylinders contain the proper amount of oil.

(6) Make sure the gun is properly lubricated (see par. 25).

(7) Check the sight for vision and see that it is clamped firmly in the periscope holder.

(8) Check the bore sight. See that all sight adjustments are locked securely.

(9) Make sure there is a sufficient amount of ammunition, that the ammunition is clean, and that it is properly stored in the tank.

(10) Be sure no one is in the path of the recoil.

b. During firing the following points should be observed:

(1) Observe the functioning of the gun as a whole in order to anticipate any failures.

(2) Observe the functioning of the recoil mechanism and report any indication of malfunctioning to the officer or noncommissioned officer in charge of the firing.

(3) Observe the functioning of the breech mechanism (automatic operation of the breechblock, cocking, and extraction). Report any malfunction or request permission to make any necessary adjustment.

(4) Keep the breech mechanism lubricated enough to insure smooth operation of the working parts.

(5) Check each round of ammunition for dirt as it is removed from the ammunition rack in anticipation of reloading the gun.

(6) Report any failure or possibility of failure of the gun to function properly.

c. After firing the following points should be observed:

(1) See that the gun is clear and turn in all live ammunition.

(2) Clean the gun thoroughly.

(a) Swab, brush, dry, and oil the bore, chamber, and breech ring.

(b) Clean, dry, and oil all other parts of the breech and firing mechanisms.

(c) Clean, dry, and oil all other exposed metal or working parts.

(3) Replace any worn or broken parts.

(4) Replenish the recoil cylinders with recoil oil if necessary.

(5) Check the traversing and elevating mechanisms for free, smooth action.

(6) Make any necessary authorized adjustments that were not made during firing.

(7) Check the functioning of the gun.

(8) Check periscope for clear vision and broken parts. Replace parts and clean where authorized.

(9) Check completeness of tool kit.

(10) Check completeness of spare parts kit, if present.

■ 24. CARE BEFORE, DURING, AND AFTER ACTUAL COMBAT.—*a.* Before going into combat the following checks should be made:

(1) Check the bore and chamber for dirt and obstructions.

(2) Check the functioning of the gun.

(*a*) Open and close the breech.

(*b*) Check the operation of the trigger mechanisms.

(*c*) Check the operation of the firing mechanism.

(*d*) Check the action of the traversing and elevating mechanisms.

(3) Check the piston rods for proper engagement with the breech ring.

(4) Check the recoil cylinders for the proper amount of oil.

(5) Check for the presence of spare periscopes and spare head assemblies.

(6) Check the periscopes for proper vision and make sure each is bore sighted accurately.

(7) Be sure the periscope is firmly seated in the periscope holder and that all sight adjustments are locked.

(8) Make sure the ammunition rack is clean and that all dirt is removed from the top of the rack.

(9) Check for the proper amount and type of ammunition. (When applicable.)

(10) Make sure the ammunition is properly placed and the fuzes are set properly. (When applicable.)

(11) If it is anticipated that dust will be encountered, make sure the outside surfaces of the breechblock and the inside surfaces of the breech ring are absolutely dry.

(12) Check completeness of the tool and spare parts kit.

(13) Report to the tank commander that the gun is ready for action. (When applicable.)

b. During actual combat the following points should be observed:

(1) Observe the action of the gun in order to anticipate any failures.

(2) Observe the opening and closing of the breech, cocking, and extraction in order to anticipate any failures. (Improper extraction may be caused by an accumulation of dust and dirt on the breechblock guides.)

(3) Observe the action of the recoil mechanism. (Recoils or counterrecoils with a slam indicate insufficient amount of recoil oil. Returns to battery unusually slow may be a result of excessive amount of recoil oil.)

(4) Keep the breechblock lubricated to insure smooth operation.

(5) Check each round of ammunition for damage or dirt as it is removed from the ammunition rack in anticipation of reloading the gun.

(6) Place defective rounds in one corner of the ammunition rack to prevent loading them by mistake.

c. After combat the following points should be observed:

(1) When time is available clean the gun thoroughly. When the time is limited clean the breech mechanism to insure proper operation.

(2) Make any necessary adjustments and replace any damaged or worn parts.

(3) Check and clean the sights thoroughly and replace any damaged parts.

(4) Replace expended ammunition and dispose of any defective rounds. (When applicable.)

(5) Replenish the oil in the recoil cylinders when necessary (a check should be made).

(6) Recheck bore sight and secure all sight adjustments.

(7) Replace all spare parts, including sights.

(8) When time is available make all checks listed in a above.

(9) Report the necessity for any major repairs. (When applicable.)

(10) Report gun is ready for action. (When applicable.)

■ 25. LUBRICATION.—a. The bore, chamber, breech ring, and breech mechanism should be oiled after each cleaning in order to protect the gun from rust and corrosion. The breech and firing mechanisms should be oiled daily when the gun is in continuous use to insure the free and smooth action of the mechanisms.

b. The barrel bearing should be lubricated every 2 or 3 days of continuous firing by means of the grease fitting on the right side of the cradle. Each week the traversing and elevating racks should be cleaned and greased. Every month the Allen setscrew should be removed from the top of the elevating gear and traversing pinion housings and the supply and the condition of the grease checked. If the grease contains grit, it should be removed and replaced by the ordnance. If the supply of the grease is low, the housing should be filled through the Allen setscrew recesses. The closing spring chain should be greased once a week. All other working parts should be oiled frequently enough to insure smooth working.

c. The traversing and elevating mechanisms should be repacked each 6 months of continuous use or when the tank is turned into the ordnance for the 500-mile check. This is an ordnance operation. The same applies to the vertical and horizontal trunnions as there are no grease fittings for the use of the using services. For type of lubricant to be used see the lubrication chart below.

LUBRICATION CHART

Parts lubricated	Frequency	Required lubricant		Amount
		Below 32° F.	Above 32° F.	
Bore	Daily	Engine oil SAE 10.	Engine oil SAE 30.	Film.
Breech mechanism	____do	SAE 10	SAE 30	Do.
Traversing and elevating racks.	Weekly	O. D. grease No. 00.		Medium coat.
Traversing and elevating pinion bearings.	Monthly	O. D. grease No. 0.		Fill housing.
Closing spring chain	Weekly	O. D. grease No. 00.		Light coat.
Barrel bearing	Every other day	____do		Several turns.
Traversing and elevating gear housings (repack).	Every 6 months or 500 hours.	O. D. grease No. 0.		Fill.
Trunnion bearings (repack).	____do	____do		Do.
All other exposed or working parts.	Daily	SAE 10	SAE 30	Film.

■ 26. Care After Subjection to Chemical Attack.—*a.* When nonpersistent gases have been encountered all parts of the gun that have been exposed to the gas should be cleaned as soon as possible to prevent corrosion. Dry-cleaning solvent, denatured alcohol, or strong soap and water may be used to clean the gun. All parts should be rinsed with water, dried thoroughly, and coated with oil.

b. When persistent gases (mustard, lewisite, etc.) have been encountered, the oiled parts of the gun should be wiped dry with a rag on the end of a stick before a decontaminating agent is used. *Whenever a persistent gas must be removed from the gun, protective clothing and a service gas mask should be worn.* Dry-cleaning solvent, kerosene, chloride of lime or the special noncorrosive decontaminating agent may be used for decontaminating the gun. The special noncorrosive agent is the most desirable agent, however, chloride of lime will serve the purpose but will cause corrosion. (Do not use powdered chloride of lime as flaming will occur when this agent is applied, in the powdered form to liquid mustard.) If chloride of lime is used mix it with an equal part of water, and if there is sufficient time allow it to remain on the gun about 2 hours. (At the present time a spray gun which will contain about 1 quart of the noncorrosive decontaminating agent is being procured, and one will be carried in each tank.)

c. After the gun is decontaminated it must be thoroughly cleaned, rinsed, dried, and oiled to insure the complete removal of all gas, corrosion, and decontaminating agent. Rags and sticks that have been used in this process should be buried as they will remain a constant source of danger to the gun crew. If the gas has penetrated the inside of the tank, the tank must be treated in the same manner.

d. Leather or canvas must be scrubbed with the bleaching solution or the available agent to insure decontamination. It may be necessary to bury all gun covers.

Section V

STOPPAGES AND IMMEDIATE ACTION

■ 27. General.—*a*. (1) A stoppage is unintentional cessation of fire caused by the malfunction of the gun or ammunition.

(2) Immediate action is the procedure used to reduce promptly such stoppage.

b. Most stoppages can be prevented by careful adherence to the instructions given in section IV.

■ 28. Stoppages.—*a*. Stoppages are generally classified as failure to fire, failure to feed, and failure to extract. With the 75-mm gun M2, the most probable types of malfunctions will be failures to fire. Those malfunctions which are not the result of broken or worn parts can generally be laid directly to carelessness and improper care of the weapon and the ammunition.

b. When the gun first fails to fire, the position of the safety lever should be checked. Due to the location of the safety lever it is very easy to set it on the "safe" position by accidently rotating it with the right shoulder. If the safety lever is in the proper position the failure of the gun to fire may be the result of the gun staying out of battery, failure of the firing mechanism, failure of the breech to close, or defective ammunition.

c. The gun may be held out of battery by an obstruction between the breech ring and the rear portion of the mount. Check for and remove any obstruction. An excessive amount of oil in the recoil cylinders will hold the gun out of battery. Elevate the breech as far as possible (9°) and carefully unscrew the top rear filler plug of the top recoil cylinder. After each few turns of the filler plug try to push the barrel assembly forward by hand. Do not entirely remove the plug as an excessive amount of oil may be forced from the cylinder by the piston if the gun suddenly starts back into battery. Allow the oil to drain until it will no longer flow from the cylinder, then tighten the filler plug. Repeat the process for the lower cylinder. After the plugs are tightened push the gun back into battery by hand. If no oil will drain from

42

the recoil cylinders, the gun may be held out of battery by an accumulation of dirt between the barrel bearing and the bronze liner. The dirt may be removed by unscrewing the piston rods from the breech ring, sliding the barrel back, and cleaning off the barrel bearing.

d. If the barrel assembly is in battery, recock and attempt to fire; if the gun still fails to fire remove the round, reload, and attempt to fire again. If the gun does not fire, the failure may be found in the firing mechanism. Check the trigger mechanism for proper operation. If the trigger mechanism seems stuck, check the alinement of the firing lever link and firing lever. The ordnance should replace or repair any damaged or broken parts. If the trigger mechanism functions properly remove the firing pin guide assembly, clean, and replace any damaged or broken parts. If the firing pin guide assembly and firing spring are not damaged, disassemble the entire breech mechanism, clean, and replace any weak or broken parts.

e. If the breech is open, check the position of the firing plunger to be sure it has not engaged the sear recess, which is just above the sear arm recess in the breechblock. If the firing plunger has made this improper engagement, jiggle the breechblock with the operating handle and pull the firing plunger from the recess with the fingers. If the breech is not held open by the firing plunger, it may be held open by a dirty or bulged round. If the round cannot be removed easily, drive the round into the chamber by using a 15- by 2- by 4-inch piece of wood as a drift against the base of the shell and to one side of the primer. An attempt to pull a badly stuck shell from the gun may leave the projectile stuck in the chamber and cause the loader to spill the powder charge inside the tank. A fused projectile stuck in the bore must be removed with a special rammer staff or by experience ordnance personnel.

f. The breech may not close if there is dirt on the surface of the breechblock. Disassemble the block and clean it. The closing spring may not be compressed enough to raise the breechblock to the closed position. Tighten the closing spring piston rod nut until the breech closes properly.

g. If the breech mechanism fails to extract the empty case, the cause may usually be traced to broken extractors. However, dirty ammunition or a dirty chamber may cause this condition. Pry or ram out the empty case, replace broken or worn parts, and clean the chamber if necessary.

h. After a complete inspection of the gun has been made and the failure of the gun to function properly has not been found, turn the gun in to the ordnance for further inspection.

■ 29. IMMEDIATE ACTION.—The following is a table of immediate action for the most usual stoppages:

Gun fails to fire

Check position of safety lever—Place in firing position

Gun still fails to fire

Not in battery	*In battery*	*Breech not closed*
Check for obstruction between breech ring and mount—Remove. No obstruction. Depress muzzle 9°. Remove oil and push to battery. Relay and fire.	Cock piece by hand and attempt to fire. Gun still fails to fire. Check firing mechanism. Gun still fails to fire. Remove round, reload, and fire.	Obstruction between breechblock and breech ring—Remove obstruction. Close breech; fire. No obstruction. Dirty or bulged round. Remove round, reload, and fire.

A quick inspection of the gun should disclose the necessary line of action to be taken. If the above chart has been followed and the gun remains out of action, a more detailed inspection must be made.

SECTION VI

SIGHTING AND SIGHTING EQUIPMENT

■ 30. GENERAL.—After insuring the proper operation of the gun, sighting becomes the all important factor of gunnery. Proper sighting cannot be overemphasized.

■ 31. EQUIPMENT.—*a.* The sight for the 75-mm tank gun is a combination periscope and telescopic sight.

b. The periscope M1 (fig. 13) is designed to be mounted in a periscope holder which follows the traversing and elevating

motion of the gun.   The periscope fits into a rectangular slot
in the periscope holder and is located by a spring loaded
plunger which drops into a hole in a block on the lower front
surface of the periscope body.   A handle is provided for withdrawing the periscope from the holder.   The periscope is a
simple mirror type, consisting of two mirrors facing each other
and inclined at an angle of 45°.   This system of mirrors
lowers the line of sight about 13 inches.   A telescope is
mounted within the periscope body in a position where its
field of view is visible through the lower mirror.   The telescope is mounted on a clip assembly so it can be readily
snapped into and out of the periscope body.

c. The reticle of the M21 telescope (fig. 16) contains a vertical crosshair and several horizontal crosshairs.   The horizontal crosshairs are based on elevations for ranges from 0 to
3,000 yards, spaced each 500 yards and numbered every 1,000
yards.   Five and ten mil lead ticks are placed on either side
of the vertical hair on the 500-yard range line.   This reticle
is designed for ammunition which has a 1,850 foot-second
muzzle velocity.

d. A battery operated instrument light is provided for arti-
ficial illumination of the reticle at night.   The cell holder
containing a single flashlight battery fits into a spring clamp
on the back of the periscope body.   The cell holder cap is
retained by pins and is removable for access to the battery.

■ 32. DISASSEMBLY AND ASSEMBLY OF PERISCOPE.—a. General.—In order for the gunner to target the gun, keep the
sight in adjustment, and maintain good visibility, it is necessary for him to be familiar with the periscope and sight.
During a single day of firing the periscope may have to be
disassembled several times to replace parts and maintain
good visibility.   Training in the disassembly of the periscope
becomes essential in the education and training of expert
gunners.

b. To remove periscope from periscope holder.—Level the
gun with the elevating handwheel in order to provide clearance between the periscope and the elevating mechanism.
Hold the spring loaded plunger, located in the lower front
of the periscope holder, out of its recess in the periscope.
Pull down on the handle which is on the bottom of the peri-

scope and slide the periscope from the holder. *Do not release the plunger until the periscope has been removed from the holder.* This will prevent the plunger from breaking the glass which protects the top mirror.

c. To remove telescope clip assembly.—Remove the instrument light assembly from the spring clamp. Insert a thin

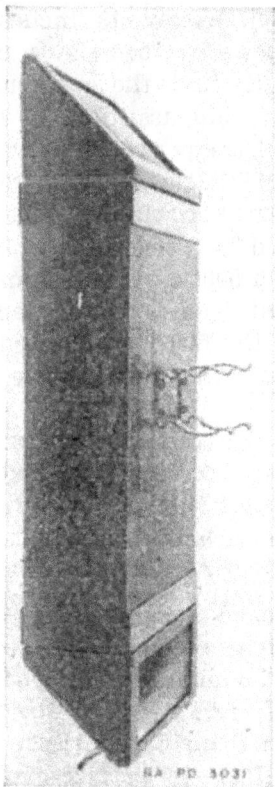

FIGURE 13.—Periscope M1.

bladed screw driver in between the clip assembly and the body of the periscope. Pry along both sides of the clip assembly, keeping the distance between the clip and periscope body the same on all sides, until the clip becomes loose. Remove the clip assembly.

d. To remove head assembly.—Squeeze the sides together at the top right hand corner of the periscope body in order

46

to disengage the dowel pins from the head assembly. At the same time remove the head assembly to the left. Normally the tank's supply of spare heads will be sufficient for combat requirements. However, in time of emergency, if it becomes necessary to disassemble the head to clean the

FIGURE 14.—Telescope with clip assembly.

mirror, the four screws that retain the inside glass may be loosened and the glass removed. Extreme care should be taken in cleaning the mirror to prevent scratching it. A clean, dry, soft cloth is necessary. Normally, the periscope will be turned in to the ordnance for repair and replacement.

47

e. To remove base assembly.—Remove the six screws, three front and three rear, at the bottom of the periscope body and remove the base assembly. In an emergency the mirror may be cleaned in the same manner as described in *d* above.

f. Assembly of periscope.—The periscope is assembled in the opposite manner in which it was disassembled. Replace

RA PD 3034

FIGURE 15.—Periscope M1—telescope and clip assembly removed.

the base assembly and tighten the six screws. Engage the head assembly with the dowel pins so that the pins will seat in their proper holes. Replace the clip assembly with the long end to the top and the sides outside the flanges of the periscope body. Snap the clip assembly into the body of the periscope. (On some types it is necessary to make sure that the instrument light cord is protruding from its recess in the clip before snapping the clip assembly to the body.)

g. To replace periscope in periscope holder.—Pull the plunger on the front face of the holder to the front and insert

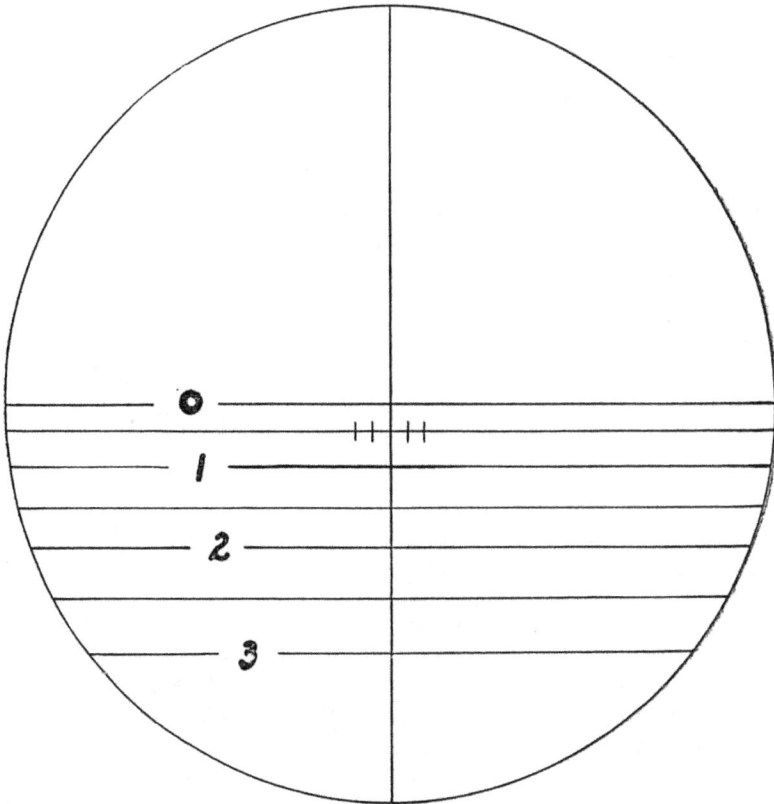

FIGURE 16.—Telescope reticle.

the periscope into the holder keeping the handle at the bottom and the glass of the head assembly to the front. Slide the periscope up in the holder until the periscope stop on the front of the periscope body strikes the bottom of the holder; *Then, and only then, release the pressure on the plunger.* Make sure the plunger seats in its recess in the periscope.

■ **33. CARE OF PERISCOPE AND SIGHT.**—*a.* It is imperative that all tanks have good, clean, and perfectly adjusted sights. During combat every target tries to make itself as inconspicuous as possible. A dirty, fogged, or scratched sight will seriously hinder and may prevent even the most experienced of gunners from delivering accurate fire.

b. The exterior surfaces of the glass should be cleaned with a soft, dry, clean cloth whenever necessary. The telescope traversing and elevating adjusting mechanisms should be oiled frequently to insure the free movement of the adjusting knobs.

c. During rainy weather the sight will be cleaned, dried, and all metal parts oiled as often as the situation will permit. Do not allow oil to be applied in such quantities that it might run onto the mirrors or lenses of the telescopic sight.

■ **34. OPERATION.**—*a.* To aim the gun in azimuth, locate the target by observing through the periscope. Pick up the target through the telescope and traverse the gun until the vertical crossline on the reticle falls on the target. Apply lead correction, if necessary, by use of lead marks on telescope reticle.

b. To aim the gun in elevation proceed as in *a* above, elevating or depressing the gun until the proper horizontal range line on the reticle falls on the target.

■ **35. SIGHT ADJUSTMENTS.**—*a. General.*—In the future all service ammunition may have similar muzzle velocities which will eliminate the difficulties encountered at the present time. However, many camps and posts will be handicapped with insufficient range facilities which will necessitate the use of reduced velocity ammunition for training purposes. In view of this fact, several methods of zeroing must be known by the using services.

b. Zeroing sight for armor piercing ammunition (1850 ft/sec velocity).—Open the breech and sight through the bore, aligning it on some distant, well-defined object. Sighting on an object 7,000 to 10,000 yards away is highly desirable. During the day, a tree on the sky line, or at night, a star that is exceptionally bright may serve the purpose.

FIGURE 17.—Instrument light for periscope M1.

Center the object in the bore by using the elevating and traversing handwheels, then without moving the barrel, by means of the adjusting knobs on the periscope, move the sight onto the target so that the intersection of the vertical crosshair and the zero range line is on the center of the object. Check the position of the bore and the crosshair, then lock the adjusting knobs by tightening the small set screws, being careful not to disturb the sight setting. The gun is now bore sighted. If the 1,000-yard range line and the vertical crosshair are placed on a target whose range is 1,000 yards, a hit can be expected.

c. Zeroing sight for reduced velocity ammunition (1,115 ft/sec velocity).—(1) The only sure way to zero the 75-mm gun, when reduced velocity ammunition is used, is to fire the gun. However, if the gunner will perform the operations in (2) below, a considerable amount of ammunition can be saved.

(2) Open the breech and sight through the barrel, from the top of the breech to the bottom of the bore at the muzzle, to the top of a target which is 1,000 yards from the gun. Without moving the barrel, move the sight onto the target so that the intersection of the vertical crosshair and the 1,000-yard range line is on the top of the target. Fire one round. Relay on the target, then move the sight onto the strike or burst. Relay, fire and adjust until a hit is obtained. The gun is now zeroed for a 1,000-yard range. If the 500-yard range line is placed on the center of a target which is 500 yards from the gun, a hit can be expected even though the reticle is designed for ammunition of 1,850 feet per second muzzle velocity. At intermediate ranges, the gunner must watch his burst and move it onto the target by changing his sight picture.

d. Changing from armor-piercing to high-explosive ammunition.—During combat when the gun has been bore sighted in order to fire armor-piercing ammunition, and there is need to fire high-explosive ammunition, the gunner must aim higher to allow for the difference in the muzzle velocities of the two types of ammunition. The gunner can do this by using a sight setting of 1,400 yards when firing on a target whose range is 1,000 yards. If the target is 500 yards away, use a sight setting of 700 yards. Gunners must interpolate for intermediate ranges.

e. Large sight adjustments.—To make large adjustments in the sight, loosen the two lock nuts on each end of the connecting bar (link) and turn the bar to raise the line of sight the desired amount. After the gun is bore sighted tighten the lock nuts. Small adjustments for differences in periscopes can be made by using the adjusting knobs on the periscope. There is no large adjustment for deflection (traverse). If the deflection adjustment cannot be made, try several different periscopes. If this does not give the desired result, inform the ordnance immediately.

f. Spare periscopes.—All spare periscopes that are carried into combat should be bore sighted for the particular gun with which they are to be used. After they are bore sighted, the adjustment knobs should be locked in place by the small set screws in the knobs. When one sight has been disabled, it can be replaced immediately by the spare, and the gun will remain bore sighted.

SECTION VII

ACCESSORIES AND SPARE PARTS

■ 36. ACCESSORIES.—Accessories include tools for such assembling and disassembling as prescribed in section II, making adjustments, and cleaning the gun. A complete list of accessories is contained in SNL K–1 and C–35. The tools for the gun are issued as part of the tool kit for the tank.

■ 37. SPARE PARTS.—SNL C–35 lists the spare parts for the gun.

SECTION VIII

AMMUNITION

■ 38. GENERAL.—Ammunition for the 75-mm gun M2 is issued in the form of fixed complete rounds. The word "fixed" signifies that the cartridge case which contains the propelling charge and primer is crimped rigidly to the projectile and that the round is loaded into the gun as a unit. The complete round includes all of the ammunition components used in the gun to fire one round.

■ 39. NOMENCLATURE.—Standard nomenclature is used herein in all references to specific items of issue.

■ 40. CLASSIFICATION.—Dependent upon the type of projectile, ammunition for the 75-mm gun M2 may be classified as armor-piercing, high-explosive, practice, and drill. The armor-piercing projectile of this caliber, as currently designed, contains a tracer but no high-explosive filler whereas the practice projectile contains an inert filler of sand. The drill cartridge is completely inert and is intended for training in service of the piece.

■ 41. FIRING TABLES.—No firing tables are provided, laying of the gun being obtained by means of graduations on the reticle of the telescope.

■ 42. IDENTIFICATION.—Ammunition, including components, is completely identified by means of the painting, marking (includes ammunition lot number), and accompanying ammunition data card. Other essential information such as weight zone and muzzle velocity may be obtained from the marking and data card. (See figs. 18 and 19 and pars. 43 to 53, incl.)

■ 43. MARK OR MODEL.—To identify a particular design, a model designation is assigned at the time the design is classified as an adopted type. This model designation becomes an essential part of the standard nomenclature and is included in the marking on the item. Prior to July 1, 1925, it was the practice to assign mark numbers. The word "Mark" abbreviated "Mk." was followed by a Roman numeral, for example: Shell, HE, Mk. I. The first modification of a model was indicated by the addition of MI to the mark number, the second by MII, etc. The present system of model designation consists of the letter M followed by an Arabic numeral. Modifications are indicated by adding the letter A and appropriate Arabic numeral. Thus, M43A1 indicates the first modification of an item for which the original designation was M43.

■ 44. Ammunition Lot Number.—When ammunition is manufactured, an ammunition lot number, which becomes an essential part of the marking, is assigned in accordance with pertinent specifications. This lot number is stamped or marked on every loaded complete round, on all packing containers, and on the accompanying ammunition data card. It is required for all purposes of record, including reports on condition, functioning, and accidents in which the ammunition is involved. To provide for the most uniform functioning, all of the rounds in any one lot of fixed ammunition are so assembled as to consist of—

   Projectiles of one lot number (one type and one weight zone).

   Fuzes, when required, of one lot number.

   Primers of one lot number.

   Propellent powder of one lot number.

Therefore, to obtain the greatest accuracy in any firing, successive rounds should be from the same ammunition lot whenever practicable.

■ 45. Ammunition Data Card.—A 5- by 8-inch card, known as an ammunition data card, is placed in each packing box with the ammunition. When required, assembling and firing instructions are printed on the reverse side of the card.

■ 46. Painting and Marking.—*a. Painting.*—All projectiles are painted to prevent rust and by means of the color to provide a ready means for identification as to type. The color scheme is as follows:

Armor-piercing:

   When inert except for   Black: marking in white.
      tracer.

   With high-explosive   Yellow: marking in black.
      filler.

High-explosive_____ Yellow: marking in black.

Chemical_____ Gray: one green band indicates nonpersistent gas; two green bands, persistent gas; one yellow band, smoke. Marking on the projectile is in the same color as the band.

Shrapnel_____ Red: marking in black.
Practice_____ Blue: marking in white. (Projec-
                          tiles may be inert or may con-
                          tain a live fuze with a spotting
                          charge of black powder.)
Dummy or drill (inert)___ Black: marking in white.

    b. Marking.—For purposes of identification, the following is marked on the components of each round of fixed ammunition:

    (1) *On projectile, except shrapnel.*

        Caliber and type of cannon in which fired.

        Kind of filler, for example, "TNT," "WP Smoke," etc.

        Mark or model of projectile.

        Weight zone marking when required.

        Lot number. Because this is ordinarily not required after the complete round has been assembled, it is stenciled below the rotating band, in which position it is covered by the neck of the cartridge case.

On shrapnel the only marking required on the projectile is that of caliber and type of cannon in which fired, and lot number.

    (2) *On cartridge case.*—Recently, changes have been made in the marking on cartridge cases. The present practice which applies to ammunition of new manufacture as well as currently renovated ammunition is shown below compared with the older practice.

    (a) *On body of cartridge case—marking in black unless otherwise indicated.**

Old marking	Present marking
"FLASHLESS" when propelling charge is of flashless (FNH) powder.	Omitted.
"REDUCED CHARGE" between two black bands indicates reduced charge; "SUPERCHARGE" in red or black indicates supercharge; absence of such markings indicates normal charge.	"REDUCED CHARGE" between two black bands indicates reduced charge; "NORMAL" below one black band indicates normal charge; "SUPER" indicates supercharge.

    *Does not apply to shell, fixed, AP, M61, supercharge, w/tracer, 75-mm gun. There is no marking on the body of the cartridge case of this round.

Old marking	Present marking
Initials of powder manufacturer, symbol of powder and lot number.	Omitted.
Caliber and type of cannon in which fired.	Omitted.
Muzzle velocity in feet per second. On normal charge rounds the muzzle velocity is inclosed in a black rectangle.	Omitted.

(b) On base of cartridge case—marking in black unless otherwise indicated.

Old marking	Present marking
Ammunition lot number. (In older lots, stamped in the metal.)	Ammunition lot number and initials of loader.
"FLASHLESS" when propelling charge is of flashless (FNH) powder.	Omitted.
Model of projectile. Absence of such model indicates shrapnel. If chemical, the kind of filler is shown.	Model of projectile.
One diametral stripe indicates normal charge; two diametral stripes at right angles indicate reduced charge; "SUPER-CHARGE" indicates supercharge.	"NORMAL" below one diametral stripe indicates normal charge; "RE-DUCED" and two diametral stripes at right angles indicate reduced charge; "SUPER" indicates supercharge.*
Caliber, type, and model of cannon in which fired (stamped in the metal).	Caliber and model of cartridge case (stamped in the metal).
Cartridge case lot number and initials of cartridge case manufacturer (stamped in the metal).	Cartridge case lot number, initials of cartridge case manufacturer and year of manufacture (stamped in the metal).

*Does not apply to shell, fixed, AP, M61, supercharge, w/tracer 75-mm gun.

(3) *On fuze* (stamped on the body).
   Type and model of fuze.
   Manufacturer's initials.
   Year of manufacture.
   Lot number.

■ **47. Weight Zone Marking.**—Because it is not practicable to manufacture projectiles (shrapnel and armor-piercing projectiles without an explosive filler excepted) within the narrow weight limits required for the desired accuracy of fire, projectiles are grouped into weight zones in order that the appropriate ballistic corrections indicated by firing tables may be applied. The weight zone of the projectile is indicated thereon by means of crosses, one, two, three, or more, dependent upon the weight of the projectile. A weight zone lighter than one cross is indicated by L. LL indicates a weight zone lighter than L. Two crosses indicate normal weight. When manufactured, shrapnel is adjusted to standard weight by addition of more or fewer balls, hence weight zone markings are not required. Similarly, armor-piercing projectiles without an explosive filler are machined to the correct weight when manufactured.

■ **48. Care, Handling, and Preservation.**—*a.* Complete rounds and ammunition components are packed in individual moisture resistant fiber containers and then in a wooden packing box or bundle. Nevertheless, since explosives are adversely affected by moisture and high temperature, due consideration should be given to the following:

(1) Do not break moisture resistant seal until ammunition is to be used.

(2) Protect the ammunition, particularly fuzes, from high temperatures, including the direct rays of the sun. More uniform firing is obtained if the rounds are at the same temperature.

b. Do not attempt to disassemble any fuze.

c. The complete round should be free of foreign matter—sand, mud, grease, etc.—before loading into the gun.

d. Do not remove protection or safety devices from fuzes until just before use.

e. Rounds and components thereof prepared for firing but not fired will be restored to their original condition and packings and appropriately marked. (See also pars. 50 and 51.) Such rounds will be used first in subsequent firings in order that stocks of opened packings may be kept at a minimum.

■ **49. Authorized Rounds.**—The ammunition authorized for use in the 75-mm gun M2 is listed below. It will be noted that the designation completely identifies the ammunition as to type and model of the projectile and the fuze, kind of propelling charge, and caliber of the gun in which the round is fired. In the nomenclature of the ammunition a suffixed statement such as "w/fuze, PD, M48" indicates the type and model of fuze assembled thereto.

Nomenclature	Approximate weight of projectile as fired (pounds)
Service ammunition	
Shell, fixed, AP, M61, supercharge, w/tracer, 75-mm gun*_____	14. 40
Shell, fixed, HE, M48, normal charge, w/fuze, PD, M48, 75-mm gun__	14. 60
Practice ammunition	
Shell, fixed, practice, sand loaded, Mk. I, 75-mm gun (with inert PDF, Mk. IV)_____	11. 85
Blank ammunition	
Ammunition, blank (1-pound charge), 75-mm gun, M1897–16–17, and 75-mm pack howitzer M1 and M1A1_____	_____
Ammunition, blank (double pellet charge), 75-mm guns, M1897–16–17, and 75-mm pack howitzer, M1 and M1A1_____	_____
Drill ammunition	
Cartridge, drill, M7, 75-mm gun, M1897–16–17_____	_____

*Burning time of tracer is 3 seconds.

■ 50. PREPARATION FOR FIRING.—The rounds listed in paragraph 49 are issued as fixed complete rounds. Each round as removed from its fiber container is ready for firing except that in the case of high-explosive rounds when delay fuze action is required, it is necessary to set the fuze DELAY—the fuze of the high-explosive round as shipped is set SUPERQUICK. (See also par. 51.)

■ 51. FUZE, PD, M48.—*a. Description.*—This fuze, assembled to the round as issued (see fig. 19) is an impact type containing two actions, superquick and delay. On the side of the fuze near the base is a slotted "setting sleeve" and two registration lines, one parallel to the axis of the fuze, the other at right angles thereto. The line parallel to the axis is marked "SQ," the other "DELAY." To set the fuze, the setting sleeve is turned so that the slot is alined with SQ or DELAY, whichever is required. The setting may be made or changed at will with a screw driver or other similar instrument any time before firing, even in the dark, by noting the position of the slot-parallel to the fuze axis for SQ at right angles thereto for DELAY. It should be noted that in this fuze, even though set SUPERQUICK, the delay action will operate should the superquick action fail to function. This fuze is classified as bore safe; that is, one in which the explosive train is so interrupted that prior to firing and while the projectile is in the bore of the gun, premature functioning of the projectile cannot occur even though the more sensitive explosive elements in the fuze should function prematurely.

NOTE.—No attempt will be made to disassemble the fuze. The only authorized operation is that of setting the fuze. Any attempt to disassemble the fuze in the field is dangerous and is prohibited except under specific direction of the Chief of Ordnance.

b. Preparation for firing.—Prior to firing, it is only necessary to set the fuze as described above and this only when delay action is required. When shipped the fuze is set superquick.

NOTE.—If for any reason the round is not fired, restore it to its original condition and packing as provided in paragraph 48.

■ 52. PACKING.—*a.* Fixed rounds of ammunition for the 75-mm gun M2 are packed in individual fiber containers and these in bundle packings of three rounds each. While the weight of the individual round varies somewhat, depending upon the type and model, the following data are considered representative for estimating weight and volume requirements:

	Weight (Lbs.)	Volume (Cu. ft.)
Complete round without packing matériel	19	
Three rounds in bundle with packing	70	1.0

b. The over-all dimensions of the packings are:

Three-round bundle—(inches)—$29\frac{1}{16}$ by 8.10 by 7.57

c. Bundles for oversea shipments are crated.

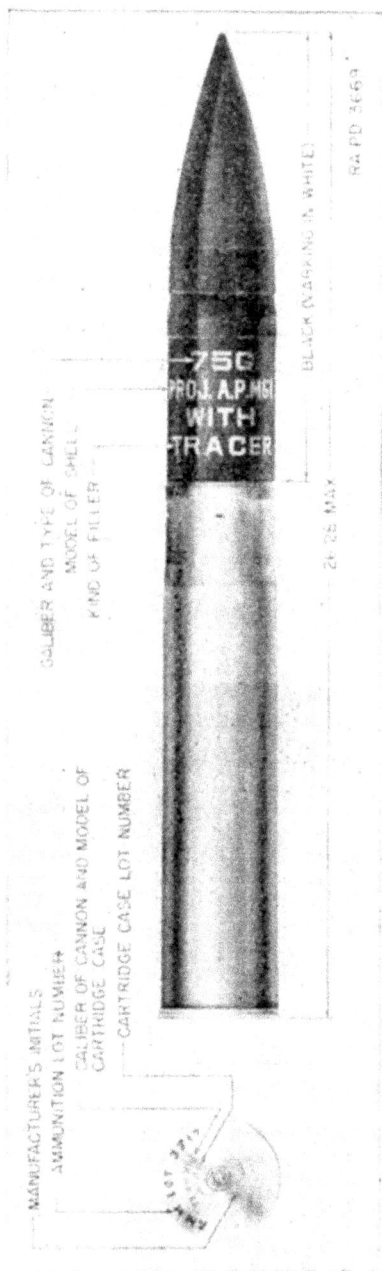

FIGURE 18.—Shell, fixed, AP M61, supercharge w/tracer, 75-mm gun.

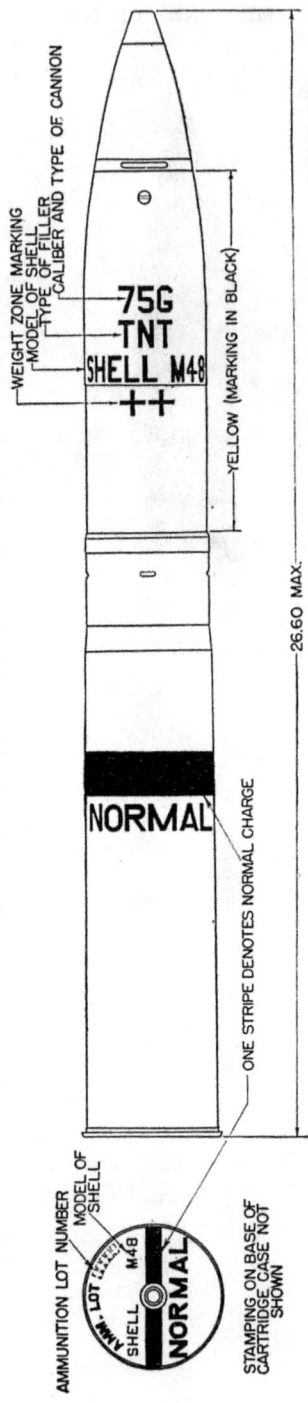

Labels on illustration:

WEIGHT ZONE MARKING
MODEL OF SHELL
TYPE OF FILLER
CALIBER AND TYPE OF CANNON

75G
TNT
SHELL M48
++

YELLOW (MARKING IN BLACK)

RA FSD 1273

26.60 MAX.

NORMAL

ONE STRIPE DENOTES NORMAL CHARGE

AMMUNITION LOT NUMBER
MODEL OF SHELL

AMM. LOT 12345

M48
SHELL
NORMAL

STAMPING ON BASE OF CARTRIDGE CASE NOT SHOWN

FIGURE 19.—Shell, fixed, HE M48, normal charge, w/fuze, PD M48, 75-mm gun.

■ 53. FIELD REPORTS OF ACCIDENTS.—Any serious malfunctions of ammunition must be promptly reported to the ord-

1. Thread recoil guide stud—lock out all recoil of carriage.
2. Trigger actuator.
3. Insert "shear block" to prevent shearing of front and rear guard screws.

FIGURE 20.—Subcaliber device.

nance officer under whose supervision the matériel is maintained for issue. (See AR 45-30.)

SECTION IX

SUBCALIBER EQUIPMENT

■ 54. 37-MM.—37-mm subcaliber equipment when developed will be issued for the 75-mm tank gun M2.

■ 55. CALIBER .22 TO CALIBER .30.—No caliber .22 to caliber .30 subcaliber equipment has been designed for the 75-mm tank gun M2. However, the equipment issued for the 37-mm guns M3 and M5 can be adapted to use in the gun. Figures 20 to

1. Shear block.
2. Trigger actuator.

FIGURE 21.—Subcaliber device.

26, inclusive, show a method of adapting this equipment. All modifications can be made locally.

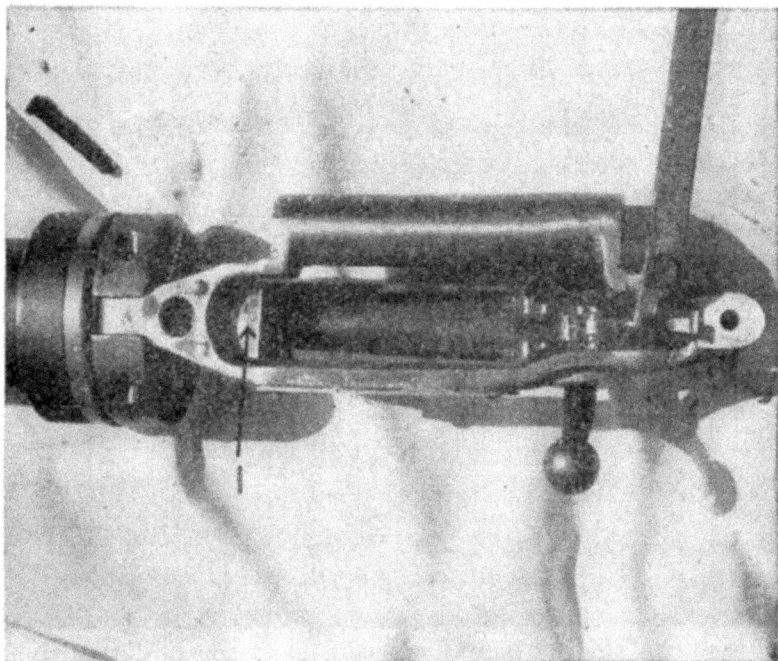

1. Shear block.

FIGURE 22.—Subcaliber.

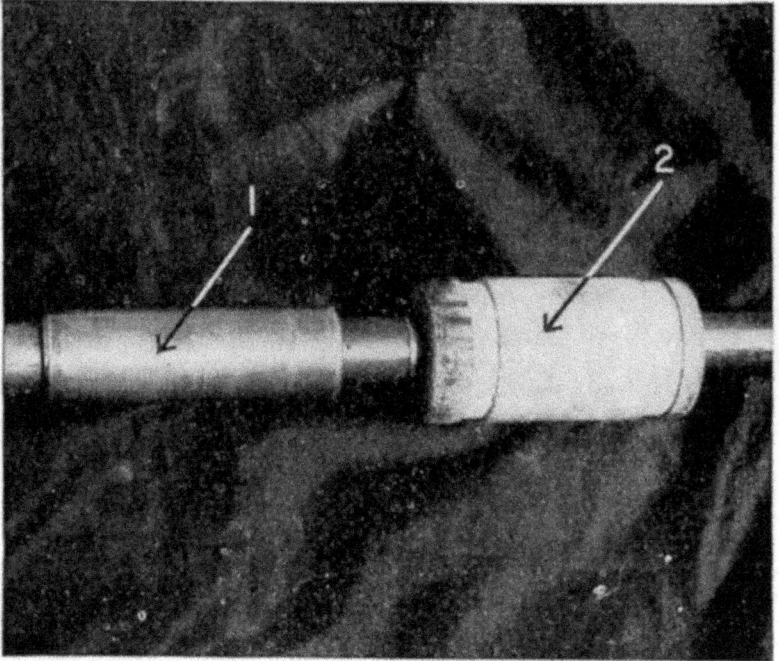

1. Extension pipe for increasing length of 37-mm liner type tube to length of 75-mm gun tube.
2. Wooden support bushing. Two are used to support the tube to prevent bending.

FIGURE 23.—Subcaliber device.

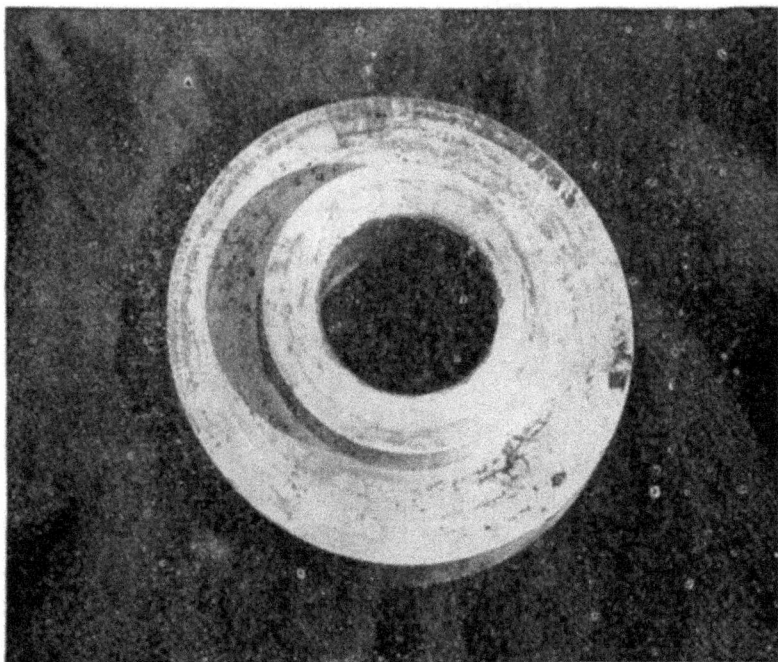

FIGURE 24.—Muzzle bushing. (Can be made quickly of wood and faced with tin or sheet iron for extra strength.)

FIGURE 25.—Subcaliber device installed in 75-mm gun. (Arrow indicates leather washer inserted to prevent damage to breech of gun.)

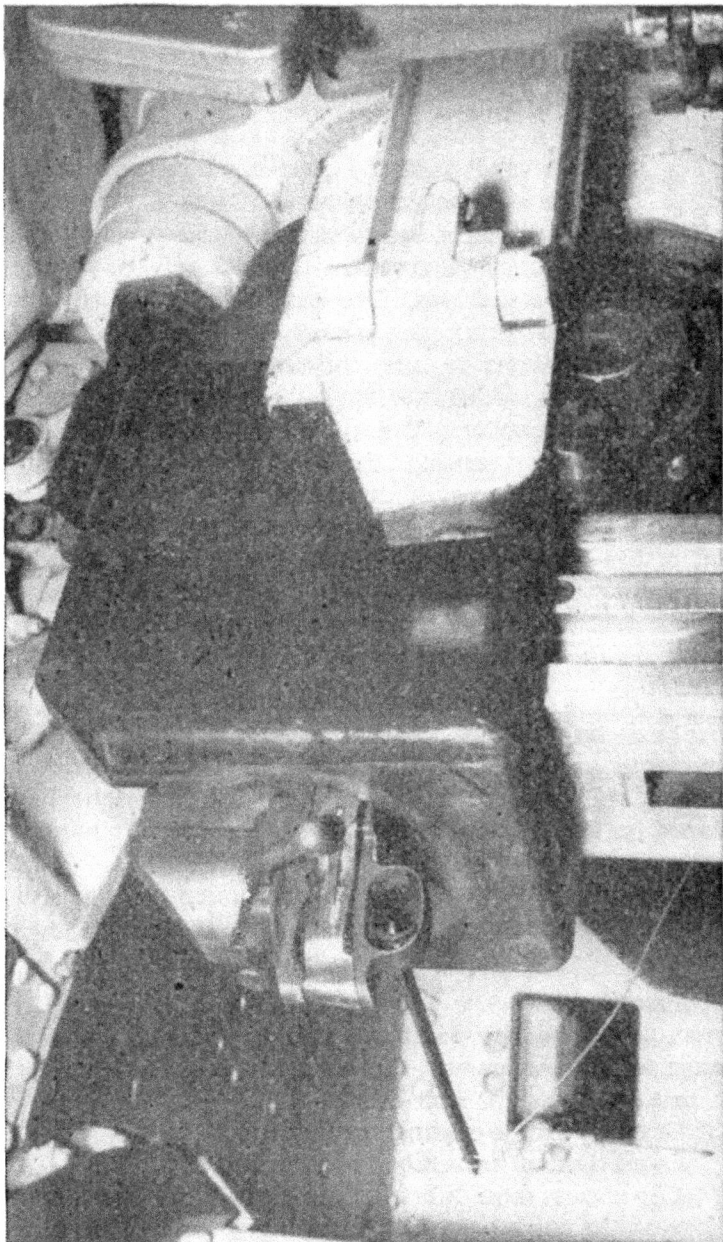

FIGURE 26.—Subcaliber device installed.

CHAPTER 2

SERVICE OF THE PIECE

■ 56. GENERAL.—*a*. Drill of the tank crew is prescribed in FM 17–5.

b. The gun crew consists of the gunner, who aims and fires the piece, and the assistant gunner, who loads the piece.

c. Training in service of the piece must stress rapidity and precision of the movement, and teamwork of the gunner, assistant gunner, and driver. The gun has a limited traverse, 14° right and left from the center, therefore the gunner must direct the driver to turn the tank as necessary.

d. Equipment is loaded into the tank under the supervision of the tank commander. The gunner is responsible for the sight and for the inspection of the gun. The assistant gunner is responsible for the ammunition.

e. When action is not imminent the gun is unloaded. When action is imminent the gun is loaded and locked; the gunner observes through the periscope; and the assistant gunner is ready to reload the new round into the gun.

f. For action in abandoning vehicle see FM 17–5.

■ 57. POSITIONS OF GUN CREW.—*a*. *Gunner*.—The gunner sits in the seat on the left of the gun and makes himself as comfortable as possible. He places his head against the sight rest and looks through the periscope; his right hand is placed on the traversing handwheel and his left hand on the elevating handwheel.

b. *Assistant gunner*.—The assistant gunner takes position in the crew compartment to the left rear of the gun out of the path of the recoil.

■ 58. PROCEDURE.—*a*. *To prepare gun for action*.—(1) The gunner, aided by the assistant gunner, checks the recoil cylinder for proper amount of oil when this is necessary and then installs the sight and checks—

　　　Operation of elevating mechanism.
　　　Operation of traversing mechanism.
　　　Cleanliness and lubrication of the gun and mount.
After receiving the report of the assistant gunner, the gunner takes his position at the gun and reports the results of his inspection to the tank commander.

(2) The assistant gunner assists the gunner in checking the recoil cylinders when that is necessary. He then, assisted by other members of the crew, loads the ammunition into the tank and inspects each round to see that it is free from dirt. He checks the tools to see that all are present and serviceable. He checks the operation of the breechblock by manually opening the breech, returning the operating handle to the vertical position, and then tripping the extractors to close the breech. He reports the results of his inspection to the gunner and takes his position.

(3) When a complete load of ammunition is to be placed in the tank, the gunner should help the assistant gunner in loading it.

b. Duties of gun crew during combat.—For care of the gun and equipment before, during, and after firing, see paragraphs 22 and 23.

(1) *Gunner.*—The mission of the gunner is to deliver accurate fire on the enemy. He must know instinctively the type of ammunition to use for each type target. The tank commander may not tell him which type to use. The designation of the ammunition to be used must be given to the assistant gunner (loader) by prearranged signals or by voice. The gunner is in command of the gun crew. During combat the gunner performs the following duties:

(*a*) Keeps on the alert to fire on targets designated by the tank commander and on any dangerous targets that he sees.

(*b*) Signals the loader the type of ammunition to be used.

(*c*) Indicates fuze setting when necessary.

(*d*) Aims the gun, and, after the loader has signaled that the gun is loaded and ready to fire, fires the gun when ready.

(*e*) Tells the gunner when to recock or reload the gun in the event of a misfire. (Checks position of safety lever before attempting to fire the second time.)

(*f*) Removes and replaces the sight when it becomes damaged.

(*g*) Determines whether or not to continue firing when either he or the loader anticipates a malfunction.

(*h*) Assisted by the loader, he reduces stoppages of the gun.

(2) *Loader.*—The mission of the loader is to keep the gun supplied with the proper type of shell and at the same time observe the operation of the various mechanisms. As the

71

gunner cannot observe the functioning of the gun and shoot at the same time, the loader must continuously observe and anticipate malfunctions. During combat the loader performs the following duties:

(*a*) Opens the breech initially when directed to do so by the gunner or when directed to load. (*If the operating handle is used, it will be immediately returned to the vertical position and latched. If the operating wrench is used, it will immediately be removed from the splined shaft and put in a convenient place.*)

(*b*) Loads the gun with the proper type ammunition when directed by the gunner. (See (*c*) below.)

(*c*) Sets the fuze on superquick or delay as directed by the gunner when high-explosive shell is used.

(*d*) Continues to load the gun with the type of ammunition that was last specified until he receives other orders from the gunner or until the supply of that type of ammunition is exhausted. (Should that type of ammunition be exhausted, the loader will inform the gunner.)

(*e*) While the round is being fired, the loader removes another round from the ammunition rack and hurriedly runs his hands over the round to remove any dirt that may have collected on it.

(*f*) Is ready to load the gun as soon as the breech opens, taking care that he is out of the path of recoil.

(*g*) In case of a misfire, unloads or recocks the gun when directed by the gunner. (If the gun is to be unloaded, the loader will open the breech, remove the round, and immediately reload unless otherwise instructed. The round that has been removed from the gun will be placed in the ammunition rack away from the other ammunition so it will not be reloaded into the gun by mistake.)

(*h*) Drives a stuck-round in the chamber by placing a 2- by 4-inch piece of wood against the base to one side of the primer and hammering on the end of the wood.

(*i*) Keeps a spare periscope near and hands it to the gunner when needed.

(*j*) Observes the functioning of the gun and immediately reports to the gunner any indication of serious malfunction.

(*k*) During lulls in firing, the loader assists the gunner in making necessary adjustments and repairs. He repairs damaged periscopes, if practicable, and checks the bore and chamber for dirt and obstructions.

c. *To load gun.*—Upon the designation by the gunner of the type of shell that will be used, the loader will load the gun. To load the gun grasp the base of the round with the right hand, withdraw it from the ammunition rack and at the same time support the nose of the projectile with the left hand. Insert the nose of the projectile into the breech opening, slide it forward over the left hand until the nose has entered the chamber and the brass case is just within the breech ring. With the heel of the right hand give the round a strong shove into the chamber, following through until the heel of the right hand is forced up from the breech ring by the breechblock. *The loader will then tap the gunner on the back with his RIGHT HAND as a signal that the gun is loaded and that he is clear of the line of recoil.*

d. *To aim gun.*—The gunner looks through the periscope, observes his target, and manipulates the traversing and elevating handwheels to aim on the target with the proper range. When the limit of traverse is reached and the gun is not yet aimed on the target, the gunner signals the driver to turn the tank by patting him on the shoulder on the side toward which movement is desired. The driver must be alert for such signals and instantly obey them.

e. *To fire gun.*—The gun being aimed on the target, the gunner, to fire the gun, presses on the firing button on the traversing handwheel. *The gunner must not fire the gun until he has been tapped by the assistant gunner signifying that the round is loaded and the assistant gunner is clear.*

f. *To unload an unfired round.*—The assistant gunner places his right hand immediately in rear of the U-shaped recess in the breech ring, opens the breech with his left hand, grasps the shell case at its base with his right hand, withdraws it to the rear. As the shell emerges from the breech ring, he grasps it near the fuze end with his left hand under the projectile, completely withdraws it from the breech ring, and returns it to its compartment.

g. *To cease or suspend firing.*—(1) *Cease firing.*—At the command or signal CEASE FIRING, the gun if loaded is unloaded. Cease firing is used to announce long pauses during firing.

(2) *Suspend firing.*—At the command SUSPEND FIRING, fir-

ing is stopped, the gun if unloaded is loaded, and is made ready for instant resumption of fire. The posts of the gun crew remain unchanged. The gunner continues to observe and lay on the target or lays on a new target if one is designated so that he may resume fire with the least practicable delay. Suspend firing is used for short pauses in firing.

(3) Formal commands for cease and suspend firing are used primarily when instructing personnel on the range.

h. To clear gun.—The gun must be cleared before anyone moves in front of the muzzle. At the command CLEAR GUN, the gun is unloaded and the breech left open. During range firing the gun will be inspected by an officer to make sure it has been cleared.

■ 59. SAFETY PRECAUTIONS.—*a.* Prior to firing, the gun will be inspected to see that there is no obstruction in the bore.

b. After firing, during range and combat practice firing, the gun will be inspected by an officer to see that it is unloaded before the tank is moved or personnel move in front of it.

c. The gunner and assistant gunner check their positions to see that no part of their bodies are in the path of recoil.

d. The gunner will not fire the piece until he has been tapped on the back by the assistant gunner to indicate that the piece is loaded.

e. Ammunition must be secured so that it will not slide around when the tank is in motion.

f. Misfires are handled as in paragraph 28. When rounds which have misfired are removed from the gun, they will be destroyed as prescribed in TM 9–1900.

g. In loading the gun, the assistant gunner will be careful not to strike the fuze of a shell against anything.

h. Metal blocks *will not* be used to drive stuck rounds into the chamber.

i. Fuzed shells will not be driven out of the gun by insertion of a staff in the muzzle, except with a staff made for that purpose.

j. Ammunition will be cleaned and placed under shelter. Do not leave it in the sun.

k. Do not attempt to disassemble any fuze.

l. When the breech is opened manually, immediately return the operating handle to the vertical position and latch it.

CHAPTER 3

MARKSMANSHIP

SECTION I

GENERAL

■ 60. PHASES OF TRAINING.—*a*. Marksmanship training is divided into three phases:

(1) Preparatory marksmanship training.

(2) Subcaliber instructions and record firing practice on the 1,000-inch moving target range.

(3) Instruction and record firing practice on the moving target field ranges.

b. The above phases are listed in order of instruction. Each man must become proficient in one phase before proceeding to the next.

■ 61. PRIOR TRAINING.—Before receiving instruction in marksmanship, the soldier must be proficient in mechanical training and service of the piece as described in chapters 1 and 2.

■ 62. METHOD OF INSTRUCTION.—*a*. The applicatory method of instruction as prescribed in FM 21–5 is used in instruction in marksmanship.

b. The platoon is the largest unit that can economically be handled for individual instruction. Explanation and demonstration may be given to groups as large as the company. The coach-and-pupil method of instruction is used to the maximum.

c. The platoon sergeant and the tank commander act as assistant instructors. The tank commander instructs his own tank crew and the platoon sergeant assists the platoon leader in supervising instruction.

d. In conducting preparatory exercises and during instruction and record firing, time is saved by having all guns operate simultaneously; the platoon leader or an instructor giving the necessary commands.

e. (1) Prior to firing on the 1,000-inch range, sights will be zeroed. Fire a shot anywhere on the target. Center a black paster over the shot hole. Without disturbing the gun, adjust the sight until the intersection of the vertical cross-hair and the 500-yard range line is on the center of the paster. Aim on another black paster and fire a shot to confirm the adjustment.

(2) For field range adjustment see paragraph 35.

SECTION II

PREPARATORY MARKSMANSHIP

■ 63. PHASES.—Preparatory exercises are divided into phases as follows:

a. 1,000-inch exercises on stationary targets.

b. 1,000-inch exercises on moving targets.

c. Long-range exercises on stationary targets.

d. Long-range exercises on moving targets.

■ 64. EQUIPMENT.—In addition to the tanks with guns installed, the following equipment is necessary for one tank platoon:

1 small portable black board.
1 portable standing frame for each two tanks.
1 target A, 1,000-inch, for each tank.
1 sled carrier.
1 stop watch.
1 tape measure, 50 feet or longer.
1 1,000-inch apparatus for operating moving targets.
1 progress and proficiency chart for each tank crew.
1 chart showing aiming pictures.

■ 65. AREA OF INSTRUCTION.—*a.* Initial preliminary marksmanship training may be held near the barracks or camp. A fairly level, cleared area approximately 75 by 200 feet will accommodate two platoons. More advanced training is conducted on the 1,000-inch moving target range. Figure 27 shows a convenient arrangement of material in the instruction area.

76

b. Group instruction is held at a convenient place in front of the tanks. The crew is then sent to the tank where individual instruction is conducted by the tank commander.

■ 66. 1,000-Inch Exercises on Stationary Targets.—*a.* These exercises may be conducted on any level piece of ground as described in paragraph 65. The 1,000-inch target as described in paragraph 86 is used.

b. The instructor first explains how to manipulate the handwheels to aim the gun. He then by means of aiming

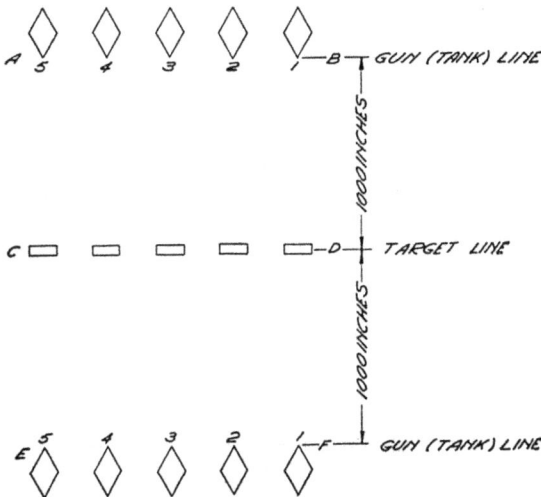

Figure 27.—Tank and tank arrangement for initial preparatory marksmanship training for two tank platoons.

charts (fig. 28), shows the proper aim when firing on stationary targets at various ranges with armor-piercing ammunition. Aiming silhouettes Nos. 1, 2, and 3 on the 1,000-inch target represent the apparent size of tanks at 300, 600, and 900 yards, respectively.

c. The gunner and assistant instructor then go to the gun. The instructor explains and demonstrates the positions of the gunner and how to manipulate the gun to aim on the target by first turning the traversing handwheel so as to bring the vertical crosshair in the center of the target and then the elevating handwheel until the proper horizontal

77

range line is on the target. He then lays the gun on the target for various ranges and causes the gunner to look through the sight. The gunner then takes his position at

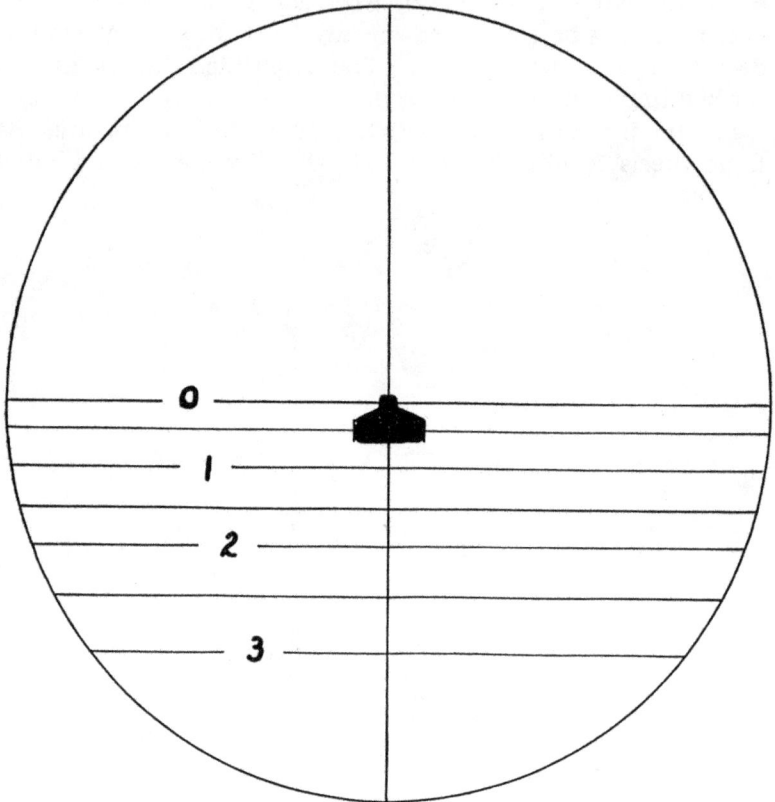

① Short range.

FIGURE 28.—Sight picture, stationary target.

the gun and lays it for various ranges as announced by the instructor. The instructor checks each laying, points out errors, and lays the gun off the target for both elevation and direction before announcing a new target.

■ 67. 1,000-INCH EXERCISES ON MOVING TARGETS.—*a. General.*— These exercises are conducted on the 1,000-inch moving target range on the 1,000-inch target. The first exercises consist of aiming for lead on a stationary target. Later exercises are conducted with targets moving.

b. *Leads.*—(1) When firing on moving targets the gun must be aimed ahead of the target to allow for the travel of the

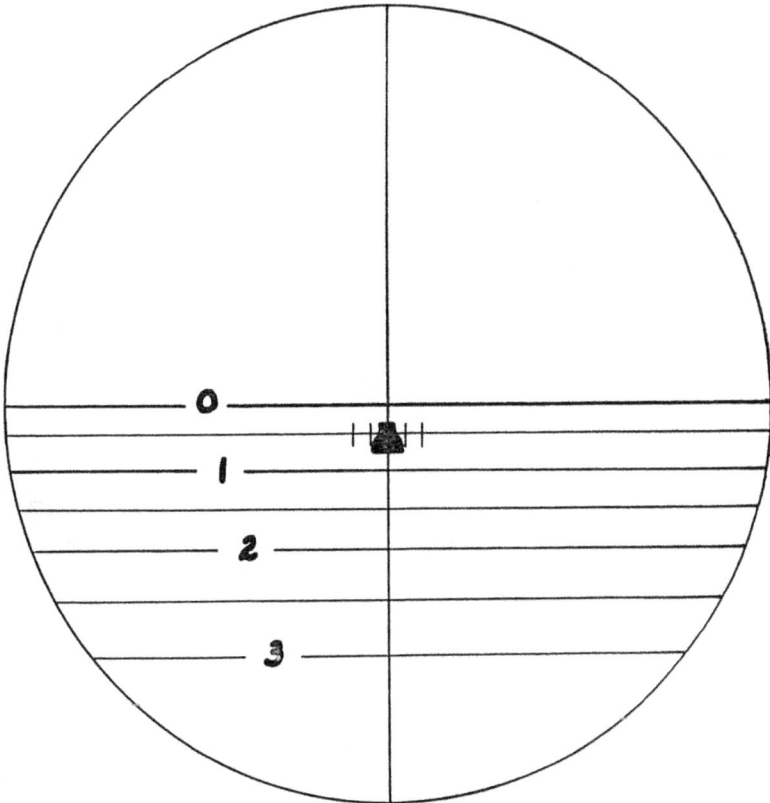

② Medium range.

FIGURE 28.—Sight picture, stationary target—Continued.

target during the time of flight of the projectile. This distance is measured in leads, one lead being one target length. Although lengths of targets will vary, the usual length of battlefield targets will be approximately 15 feet.

(2) Lead is a function of both speed and distance. It is not practicable to estimate range and speed to the last yard

or mile. Computation would be a slow process under such
conditions. Range is estimated as short (100–400 yards),
medium (400–700 yards), and long (700–1,000 yards). Speed
of the target is estimated as slow (less than 10 mph), me-

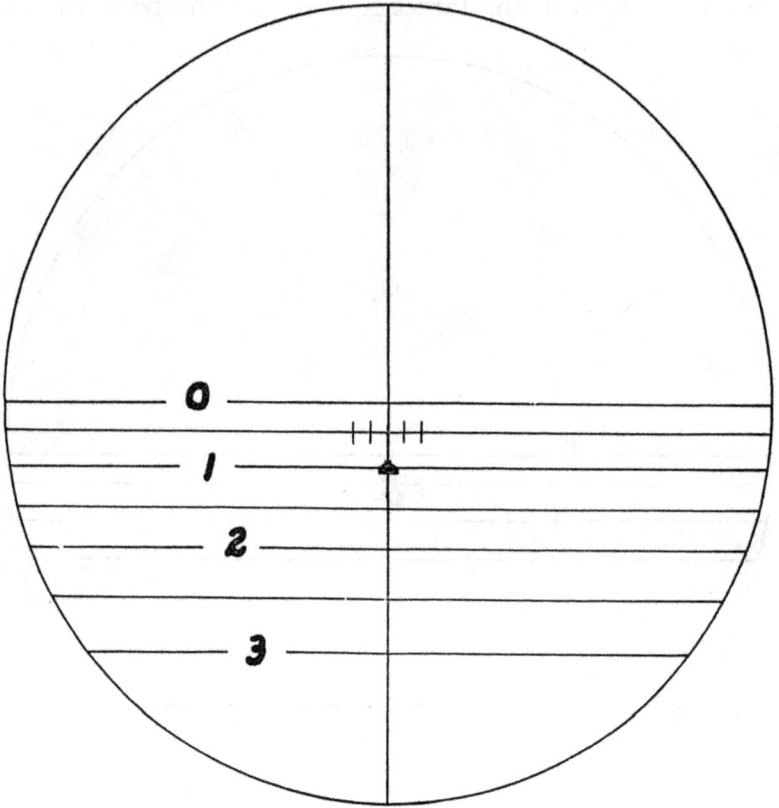

③ Long range.

FIGURE 28.—Sight picture, stationary target—Continued.

dium (10 to 20 mph), and fast (over 20 mph). Using these
factors the following lead table has been computed for the
first shot. Corrections are made upon observation of the
strike or tracer.

LEAD TABLE

Target moving at right angles to line of fire, speed mph	Short 100–400 yards	Medium 400–700 yards	Long 700–1,000 yards
Slow, less than 10	½	1	1½
Medium, 10 to 20	1	2	3
Fast, over 20	2	3	4

(a) The angle at which the target is moving with respect to the line gun-target will affect the amount of lead taken. If the angle between this line and the line of travel of the target is less than 45°, use one-half the lead shown in the table.

(b) When the target is moving directly toward or away from the gun, no lead is taken except when such target is moving up or down a steep slope.

(c) When firing from moving tanks at stationary targets, rear leads (aiming to the rear of the target) are taken in accordance with the speed of the tank. When firing from moving tanks at moving targets the sum or difference of speed is used in estimating lead.

(d) Too much lead is better than too little as the target runs into the fire and observation of the strike is easier.

(3) For initial aim at a moving target, lay the vertical crosshair of the sight on the rear of the target with the proper horizontal range line on the target. Swing the gun through the target and on in front the required number of leads. When this point is reached traverse uniformly with the target and fire. Observe the strike, estimate corrections, and again aim and fire.

c. Aiming for lead.—The instructor explains leads as in b above and by means of charts (figs. 29 to 31, incl.) indicates the proper aim for leading moving targets at the various ranges and speeds and the method of obtaining leads. The gunner and assistant instructor go to the gun. The assistant instructor demonstrates the method of aiming for lead (fig. 32) and lays the gun for leads for various ranges and speeds. The gunner looks through the sights to see the correct aim.

The gunner then takes his position at the gun and lays the gun for various leads and ranges as announced by the instructor. The instructor will vary his announcement by giving range and speed. The gunner will then lay on the target with the correct lead. After each aiming the instructor

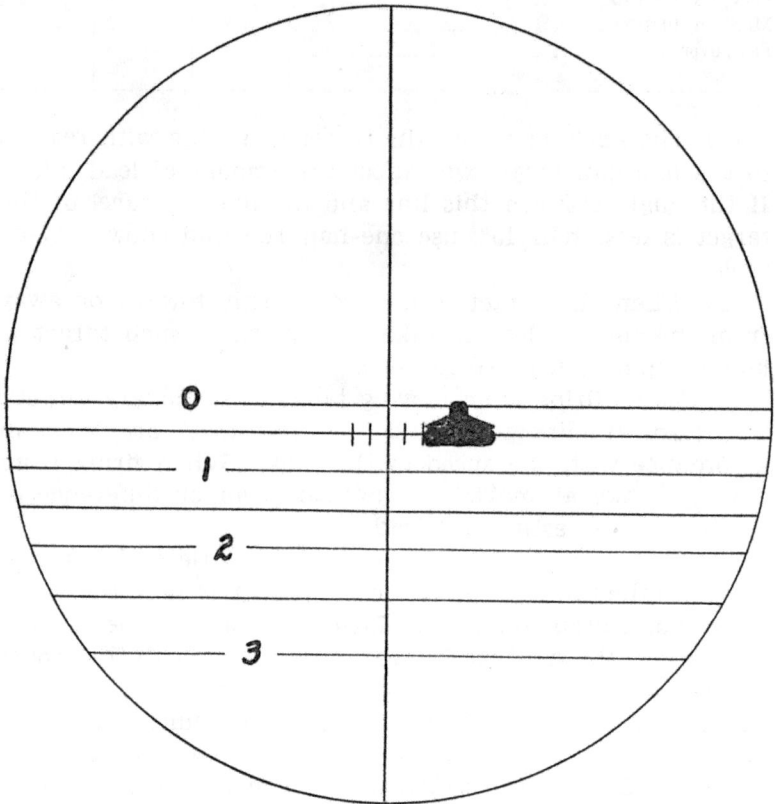

① Slow speed.

FIGURE 29.—Sight picture, moving target, short range.

checks the aim and lays the gun off the target before announcing a new range and speed.

d. *Leading a moving target on the parallel-level and parallel-hilly 1,000-inch courses.*—(1) For these exercises an organization as follows is recommended:

(a) *Platoon leader.*—Conducts instruction and supervises generally the work of the entire platoon.

(b) *Platoon sergeant.*—Issues the orders for conducting the exercises and controls, by signal, the operation of the target.

(c) *Tank commanders.*—Supervise the work in their tank;

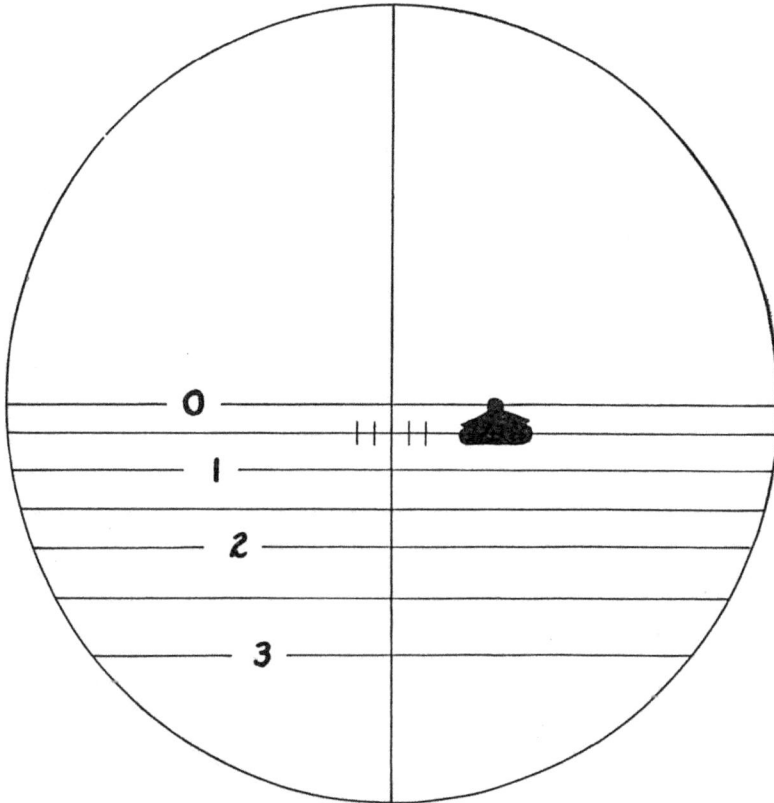

② Medium speed.

FIGURE 29.—Sight picture, moving target, short range—Continued.

relay orders from the platoon sergeant to the crew. Signal the platoon sergeant "Ready" when the gun in their tank is ready to engage the target.

(d) *One assistant instructor—timekeeper.*—Starts target on signal from platoon sergeant and regulates time of exposure in accordance with his orders. Specifies to the men operating the drum the time of exposure of the target for each run. At intervals, as an aid to regulating the target speed, calls out the time consumed as the target travels across the course.

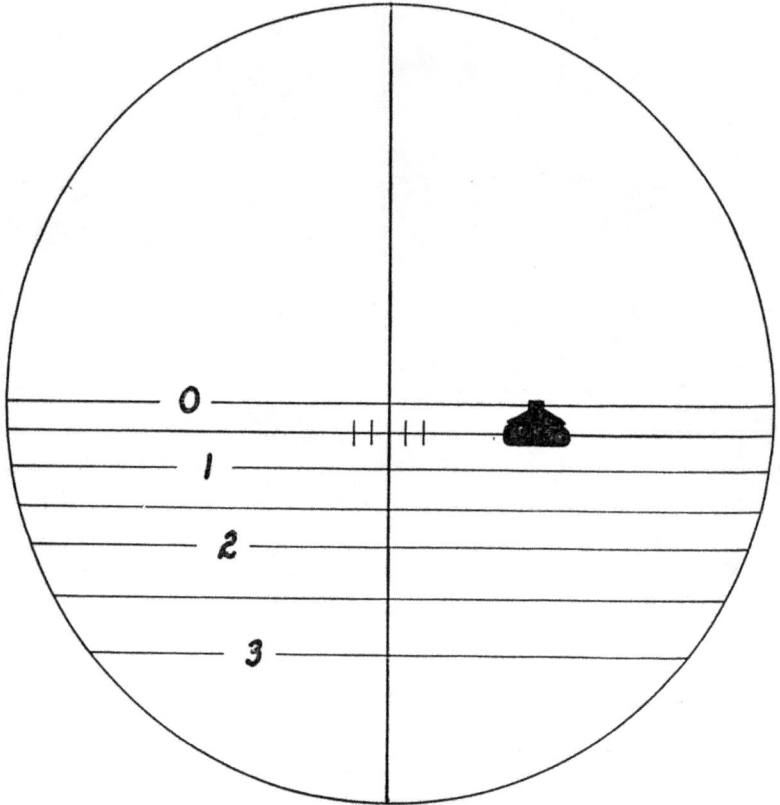

③ Fast speed.

FIGURE 29.—Sight picture, moving target, short range—Continued.

(e) *Two drum operators.*—Operate the drum so that the target will be exposed as nearly as possible for the time specified by the assistant instructor. Regulate the speed at which they turn the drum so as to obtain a uniform rate of travel for the target throughout its entire course.

(f) *Coaches.*—At the guns. Conduct individual instruction, check execution of the exercises by the gunners.

(g) *Gunners.*—Execute the exercises at the guns. The loaders assist in coaching as directed by the tank commander.

(h) *Remainder of platoon.*—Held well clear of the tanks,

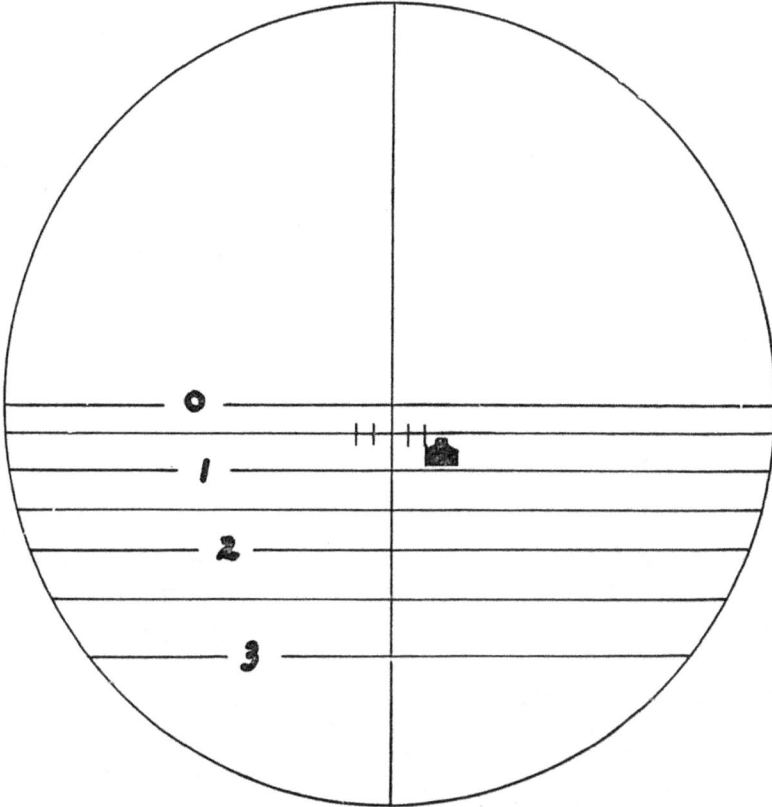

① Slow speed, medium range.

FIGURE 30.—Sight picture, moving target.

arranged numerically, ready to move forward to take their turn at the guns when so directed.

(2) The following table shows the approximate speed at which 1,000-inch moving targets should be run to represent speeds of various ranges:

TARGET SPEEDS, 1,000-INCH MOVING TARGETS

Target speeds in mph	Target speeds in inches per second corresponding to—		
	400 yards	600 yards	900 yards
10	12	8	5
15	19	12	8
20	24	16	11

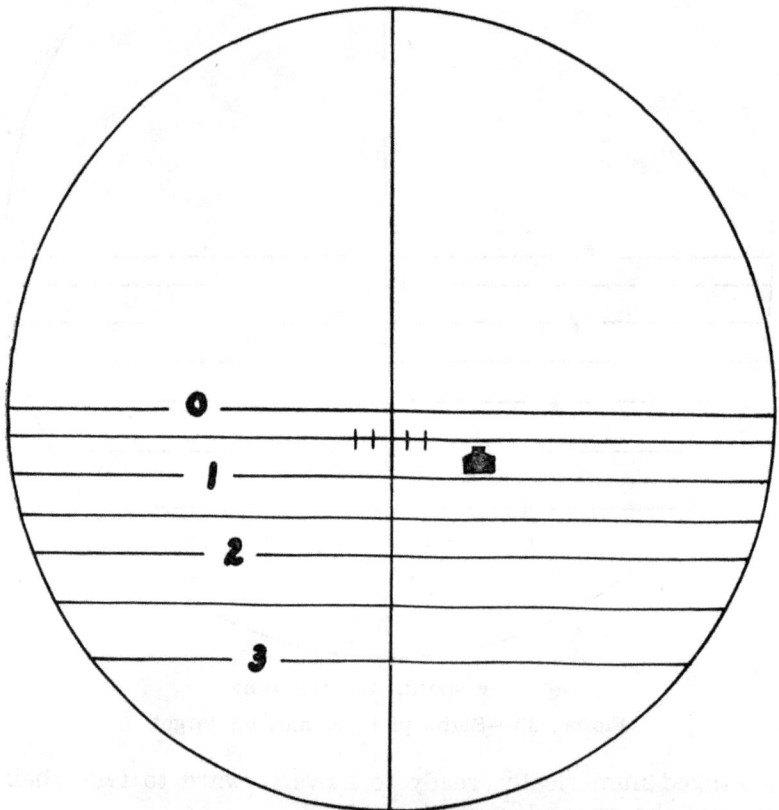

② Medium speed, medium range.

FIGURE 30.—Sight picture, moving target—Continued.

The time of exposure of a target for a particular run may be determined by dividing the distance to be traveled in inches by the target speed in inches per second shown in the table.

(3) Initially, in conducting these exercises, the slower speeds should be used; as instruction progresses, the speeds

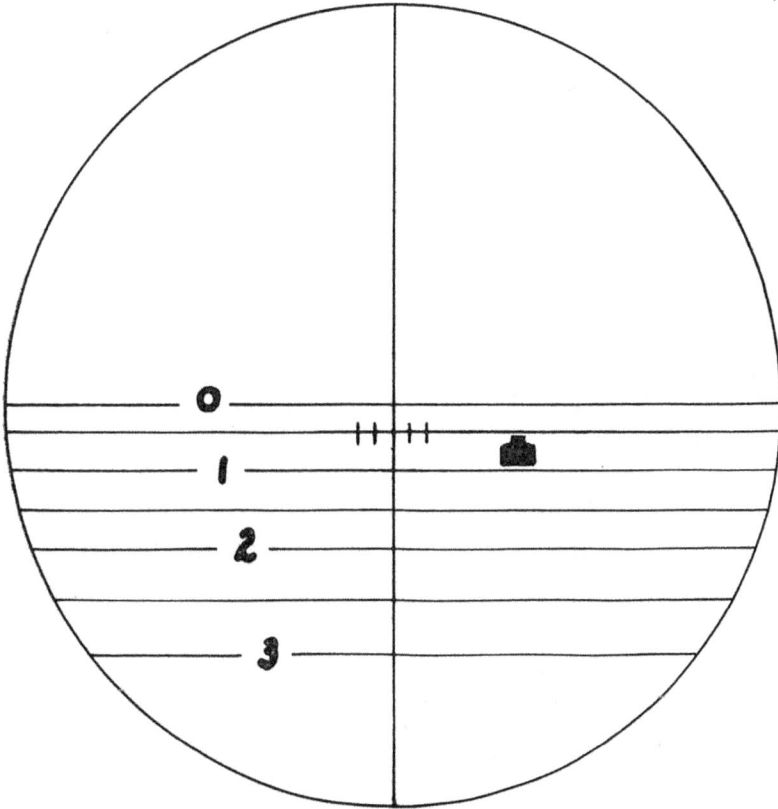

③ Medium speed, long range.

FIGURE 30.—Sight picture, moving target—Continued.

used for successive runs of the targets should be varied and increased, and finally the target should be moved at erratic speeds.

(4) The first exercises should be held with the target moving slowly and later exercises with increased speeds.

Emphasis is placed upon smoothness of operation and accuracy of laying.

(5) Each exercise is begun by a command such as LEFT FRONT, UPPER SILHOUETTE, 500, TRACK. At the command LEFT FRONT, the gunner traverses the gun to the left and when the

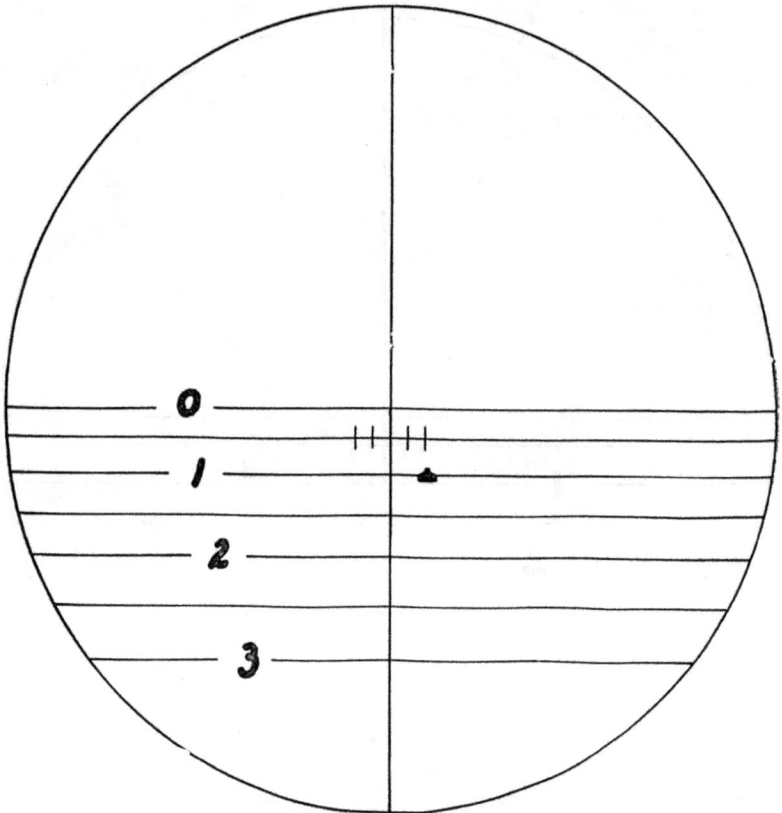

① Slow speed.

FIGURE 31.—Sight picture, moving target, long range.

target appears, aims with the announced range and lead. When the target disappears behind the screen on the right of the range, the gunner keeps his gun pointed at the point of disappearance. When the target reappears the gunner immediately begins tracking. As the gunner develops facility in tracking he is required to simulate firing.

(6) Traverse of the target across the range in one direction is known as a run and traverse across and back is known as a double run. No commands are given for traversing the target back across the range.

② Medium speed.

FIGURE 31.—Sight picture, moving target, long range—Continued.

(7) The *first exercise* is conducted on the parallel-level 1,000-inch course. The gunner takes his position at the gun and when the target appears aims with the prescribed lead and range and tracks the target. Initially only the 400 to 700 range zone is used. Later other range zones are included.

(8) The *second exercise* is conducted on the parallel-hilly 1,000-inch range course. The procedure is the same as described in (7) above. The instructor explains that the target must be lead both horizontally and vertically, that is,

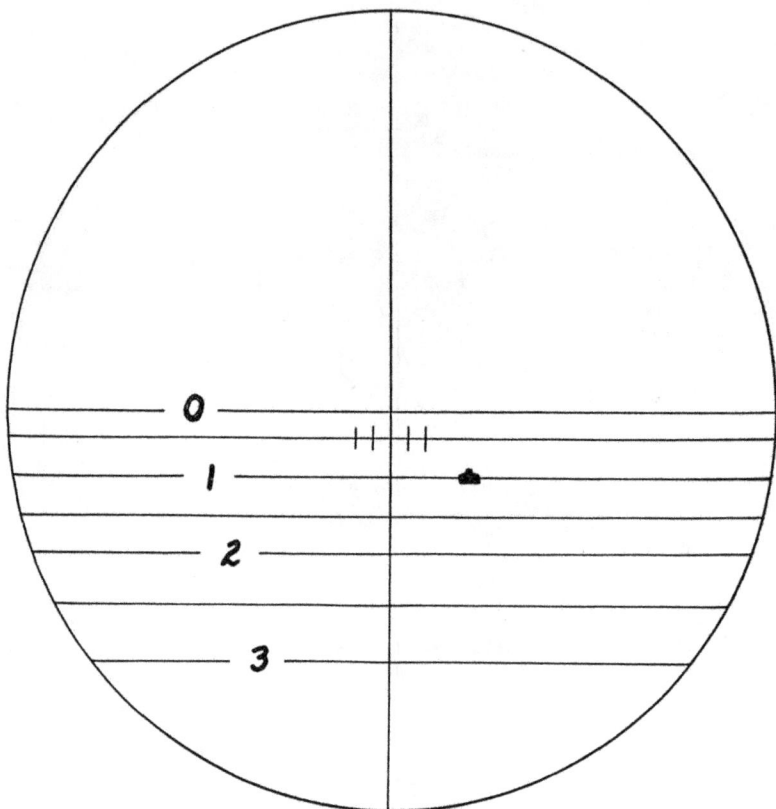

③ Fast speed.

FIGURE 31.—Sight picture, moving target, long range.—Continued.

the point of aim must be on the axis of motion. When the target reaches its highest point and then starts down, the gunner must continue traversing and at the same time rapidly depress the gun to put the point of aim again on the axis of motion. When the target reaches its lowest point this process is reversed. Figure 33 shows a method of tracking targets on the parallel-hilly course.

■ 68. LONG RANGE EXERCISES ON FIELD TARGETS.—*a.* These exercises are conducted on the field ranges. Use a field

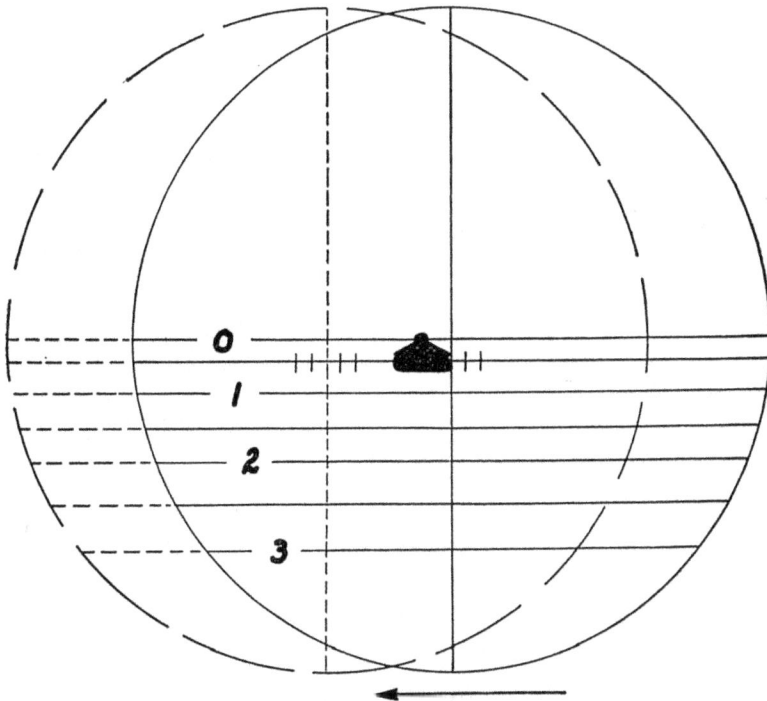

FIGURE 32.—Method of laying on moving target.

range target as explained in paragraph 86; set targets at various ranges within the three range zones—short, medium, and long.

b. Have targets between 700 and 1,000 yards from the gun, course them, conduct first exercises on stationary targets and, later, on moving targets as in paragraph 67.

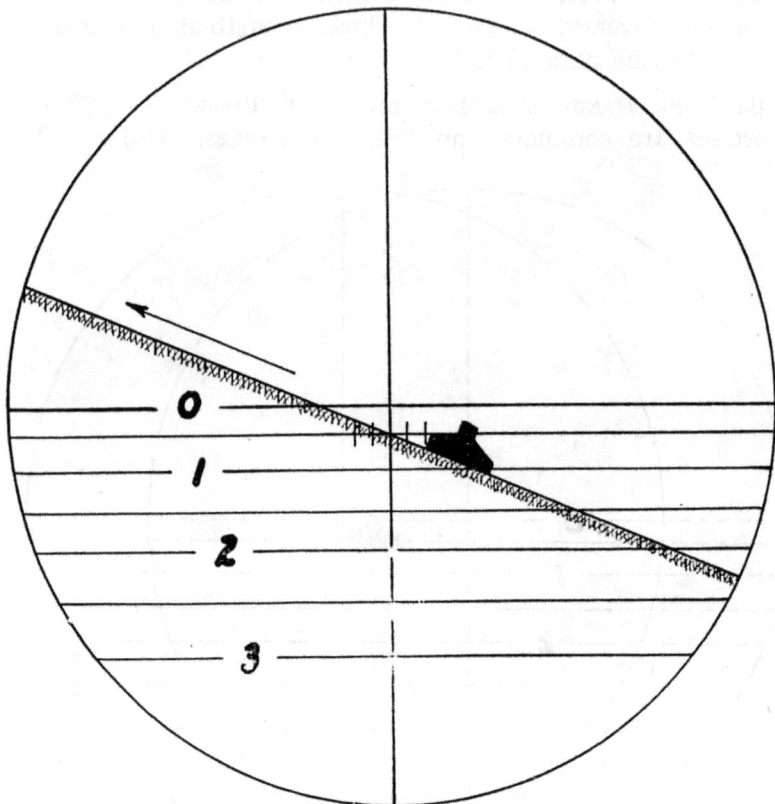

FIGURE 33.—Method of tracking target up or down hill.

SECTION III

RANGE PRACTICE

■ 69. GENERAL.—*a.* The gunner and the assistant gunner of the 75-mm tank gun M2 mounted in the medium tank M3 will fire one of the courses given below for qualification. The course to be fired will be designated by higher headquarters. Ammunition allowances and qualification scores are prescribed in AR 775–10; records and reports in AR 345–1000; and compensation in AR 35–2380.

b. Prior to firing, all men will complete the preparatory marksmanship course as prescribed in section II.

c. All firing on the 1,000-inch range will be executed with an assumed range of 600 yards. Caliber .22 ammunition will be used, except on field ranges.

■ **70. COURSE A.**—Ranges: as indicated under each table. Targets: for tables I and II, target A, silhouette as indicated; for table III, target B.

a. Instruction practice.—Fire tables I, II, and III three times.

NOTE.—Instruction and record practice firing of tables I and II will be completed before proceeding to table III.

TABLE I.—*Parallel-level course*

(Stationary tank—moving target)

Range	Silhouette number	Number of rounds	Speed in inches per second	Time of traverse in seconds	Lead	Direction of move- ment
1,000-inch_____	1	5	8	44	0	L. to R.
Do_____	1	5	8	44	0	R. to L.
Do_____	2	5	12	30	1	L. to R.
Do_____	2	5	12	30	1	R. to L.
Do_____	3	5	16	22	2	L. to R.
Do_____	3	5	16	22	2	R. to L.

TABLE II.—*Parallel-hilly course*

(Stationary tank—moving target)

Range	Silhouette number	Number of rounds	Speed in inches per second	Time of traverse in seconds	Lead	Direction of move- ment
1,000-inch_ ___	1	5	8	44	0	L. to R.
Do_____	1	5	8	44	0	R. to L.
Do_____	2	5	12	30	1	L. to R.
Do_____	2	5	12	30	1	R. to L.
Do_____	3	5	16	22	2	L. to R.
Do_____	3	5	16	22	2	R. to L.

TABLE III [1]

Range	Number of rounds	Speed	Lead	Direction of movemen of target
Field range (not less than 700 yards).	10	Uniform speed of approximately 15 mph.	As necessary.	L. to R. and back to 1 Exposed 15 secon(each direction. Gu: ner fires one roun for each exposure.

[1] Caliber .30 tracer ammunition is used in this firing.

b. Record practice.—Fires tables I, II, and III of cours A once under prescribed record firing conditions.

■ 71. COURSE B.—Range: 1,000 inches. Targets: target A silhouette as indicated.

a. Instruction practice.—Fire tables IV and V three times

TABLE IV.—*Parallel-level course*

(Stationary tank—moving target)

Range	Silhouette number	Number of rounds	Speed in inches per second	Time of traverse in seconds	Lead	Direction of move ment
1,000 inch	1	5	8	44	0	L. to R.
Do	1	5	8	44	0	R. to L.
Do	2	5	12	30	1	L. to R.
Do	2	5	12	30	1	R. to L.

TABLE V.—*Parallel-hilly course*

(Stationary tank—moving target)

Range	Silhouette number	Number of rounds	Speed in inches per second	Time of traverse in seconds	Lead	Direction of move ment
1,000-inch	1	5	8	44	0	L. to R.
Do	1	5	8	44	0	R. to L.
Do	2	5	12	30	1	L. to R.
Do	2	5	12	30	1	R. to L.

b. Record practice.—Fire tables IV and V of course **B** once under prescribed record firing conditions.

■ 72. COURSE C.—Range: 1,000 inches. Target: target **A**.
a. Instruction practice.—Fire tables VI and VII three times.

TABLE VI.—*Parallel-level course*

(Stationary tank—moving target)

Range	Silhouette number	Number of rounds	Speed in inches per second	Time of traverse in seconds	Lead	Direction of movement
1,000 inch_____	2	5	12	30	1	L. to R.
Do_____	2	5	12	30	1	R. to L.

TABLE VII.—*Parallel-hilly course*

(Stationary tank—moving target)

Range	Silhouette number	Number of rounds	Speed in inches per second	Time of traverse in seconds	Lead	Direction of movement
1,000-inch_____	2	5	12	30	1	L. to R.
Do_____	2	5	12	30	1	R. to L.

b. Record practice.—Fire tables VI and VII of course C once under prescribed record firing conditions.

SECTION IV

CONDUCT OF RANGE PRACTICE, INCLUDING RECORD PRACTICE AND INDIVIDUAL SAFETY PRECAUTIONS

■ 73. GENERAL.—Rules and regulations governing range practice are prescribed herein to secure uniformity throughout the service.

■ 74. DUTIES OF PERSONNEL.—*a. Officer in charge.*—The officer in charge of range practice, detailed by the unit commander, is responsible for—

(1) Assignment, coordination, and supervision of ranges and firing areas.

(2) Timely arrangements with the range officer for repairs or alterations of installations.

(3) Procurement of supplies for firing units.

(4) Enforcement by firing units of safety precautions prescribed herein, in AR 750–10, and by the local commander.

(5) For interpretation of such parts of this manual as may be referred to him for decision.

b. Range officer.—The range officer is normally a member of the unit or post commander's staff. He is responsible for—

(1) Procurement and distribution of range supplies.

(2) Supervision of construction, alteration, or repair of range installations.

(3) Establishment of safety limits of ranges and coordination of firing to comply with the provisions of AR 750–10.

c. Company commander.—(1) The company commander is responsible for the efficiency of the marksmanship training of his organization and the conduct of its firing in accordance with the provisions of this manual.

(2) During 1,000-inch target firing he or his commissioned representative will personally supervise and control all firing by one of the following methods:

(*a*) Give the commands COMMENCE FIRING and CEASE FIRING for each order.

(*b*) Give the command COMMENCE FIRING, then permit tanks to fire individually, observing the safety precautions prescribed in paragraph 75, until he gives the command CEASE FIRING.

(3) During moving target firing, he or his commissioned representative will control the firing and the movement of the target(s) by appropriate fire orders and signals. A typical order would be: LOAD, 1. RANGE 500, RIGHT ONE LEAD, UPPER SILHOUETTE, 2. COMMENCE FIRING.

(4) During all firing, he or his commissioned representative will rigidly enforce local range regulations, safety precautions as prescribed in paragraph 75, and instructions pertaining to the service of the piece contained in chapter 2.

d. Scoring officers.—(1) Scoring officers will be detailed to supervise record firing practice. Officers for this duty will be detailed from organizations other than the one firing. They will familiarize themselves thoroughly with their duties and firing procedure on the 1,000-inch and field firing ranges prior to the date of commencement of record firing practice. The number of scoring officers detailed during record firing practice will not be less than one for each four tanks firing (or for each 1,000-inch range unit being operated in the case of 1,000-inch moving targets). An assistant to the scoring officer will be present in each tank during the firing.

(2) Specific duties of the scoring officers are to—

(a) Inspect loaded magazines and count number of rounds of ammunition to be fired by the gunner for each exercise. (This may be delegated to the assistant scorer.)

(b) Check dimensions of the targets, aiming silhouettes, and scoring spaces, and see that the range is laid out as prescribed.

(c) See that firing is conducted in accordance with the prescribed procedure.

(d) Verify and render decisions on all misfires, stoppages, and malfunctions of the guns. (See par. 82.)

(e) Render a decision in event of breakage or stoppages in any of the range apparatus or mechanism.

(f) Inspect each target before it is placed on the sled. Make sure that initially it contains no shot holes and that after being fired on it has no unpasted shot holes before the start of another score.

(g) Count the number of shot holes in appropriate scoring space for each score fired, score target(s), and record the score.

(h) Check time of exposure of target on each run and render a decision in event of irregularities.

(3) Additional specific duties of assistant to the scoring officers are to—

(a) Count ammunition when so directed.

(b) Relay commands to gun crew.

(c) During moving target firing, give the commands or signals for COMMENCE FIRING and CEASE FIRING when target enters and leaves the zone for firing.

(d) See that gunner executes firing in accordance with prescribed procedure.

(e) Report all stoppages and discrepancies to scoring officer for decision.

(*f*) See that gun is properly cleared at completion of firing.

e. Coaches.—(1) *General.*—(*a*) *During instruction firing.*—During all preparatory training and instruction firing there will be a coach at each firing gun (tank). It is his task to see that the gun crew executes all operations in accordance with the prescribed procedure. His main duty is to detect errors and cause the gun crew to correct them.

(*b*) *During record firing practice.*—The gunner will not be coached or instructed in any way except as in paragraph 85. Coaches will not be allowed at or near the guns.

(2) *Specific duties.*—(*a*) Require each man functioning at his gun to observe all pertinent individual and general safety precautions as prescribed in paragraph 81, and see that the instructions pertaining to the service of the piece contained in chapter 2 are complied with.

(*b*) See that the proper amount of ammunition is at the gun and that magazines are loaded with the specified number of rounds for each exercise.

(*c*) Supervise generally the work at the gun, making sure that the commands LOAD, COMMENCE FIRING, CEASE FIRING, UNLOAD, and CLEAR GUN are properly executed. He will repeat orders or instructions when necessary to insure correct understanding and timely execution by the gunner and loader.

(*d*) See that the gunner executes the firing exercises in accordance with prescribed procedure.

(*e*) Report all misfires, stoppages, malfunctions, or discrepancies to the officer conducting firing.

(*f*) Score the target when directed and discuss the execution of the exercise (during instruction practice only) with the gunner.

f. Gunner.—The gunner will fire the prescribed tables in accordance with the procedure given in paragraph 80.

g. Loader.—(1) *General.*—The primary duty of the loader is to serve the piece in that capacity during all firing exercises. During instruction practice he may also perform additional duties at the gun or act as an assistant coach in accordance with orders of the coach. During record firing practice, he performs only the specific duties of loader hereinafter prescribed; he does not coach or instruct the gunner in any way.

(2) *Specific duties.*—(a) Secure and have ready for use at the gun the prescribed number of rounds for each exercise.

(b) Serve the piece as loader as prescribed in chapter 2. In this connection he will—

1. Load the gun in accordance with the commands of the officer conducting the firing and tap the gunner on the back when the gun is loaded.

2. When the gun is loaded and ready to fire and when the gunner is ready to begin the exercise, signal "Ready" to the officer conducting the firing.

3. Repeat all orders to unload, cease and suspend firing, and clear gun, and see that the orders are complied with as described in chapter 2.

4. Announce for the gunner's benefit when the prescribed number of rounds for each run of the target have been fired, thus "Five rounds complete."

(c) Report all misfires and malfunctions or stoppages to the coach (or to the scoring officer during record firing practice). During instruction firing, in case of a misfire immediately reload without command and endeavor to render the gun ready for firing as quickly as practicable. In case of a misfire, stoppage, or minor malfunction during record firing practice, proceed as described in paragraph 82.

(d) Without command, reload the gun upon the completion of a run of the target.

■ 75. SAFETY PRECAUTIONS.—*a. Responsibility.*—The responsibility and duties of individuals incident to the observation and enforcement of safety precautions are given in paragraph 74.

b. Ranges.—The specifications for laying out ranges to comply with safety requirements for the various types of firing are given in AR 750–10.

c. General and individual safety precautions.—The safety measures to be observed while firing ammunition in peacetime are laid down in AR 750–10; however, the following precautions are given for emphasis:

(1) Danger flags will be displayed at prominent positions on the range during the firing.

(2) Such range guards as may be needed will be posted.

(3) During moving target firing, conspicuous markers will be placed to indicate the right and left safety limits of the range. All concerned will be instructed in the significance of these markers.

(4) Firing on any range will not commence until it has been determined that the range is clear and that the officer in charge of firing has given his authority to fire.

(5) No firing will be done except under the direct supervision of an officer.

(6) No gun will be loaded until a command to do so has been given.

(7) No person will be allowed in front of a tank for any purpose until so directed by an officer or noncommissioned officer, who has ascertained that all guns of the tank are cleared.

(8) No gun (tank) will be removed from the firing point until an officer has inspected the guns to see that they are unloaded.

(9) A scarlet flag will be displayed on each tank from which firing is being conducted. This flag will not be removed until all guns in the tank concerned have been cleared.

(10) Decision as to whether it will be necessary to have flank tanks clear guns to permit adjacent tanks to change targets will be made by the officer in charge of range practice.

■ 76. GUNS, MOUNTS, AND SIGHTS.—a. The gun, mount, and sight will be used as issued by the Ordnance Department without addition or modification except as specifically authorized hereafter.

b. Before marksmanship firing is begun, each gun, mount, sight, and accessories to be used will be thoroughly examined and repaired or adjusted to insure their efficient functioning. Excessive play will be removed from the guns and mounts and adjustments made to permit smooth manipulation.

■ 77. TARGETING.—a. Prior to firing, the alinement of the sight on each gun will be verified and if necessary adjusted as described in paragraph 62. The sight with which the gun is to be equipped during marksmanship firing will be used for targeting.

b. Guns will be targeted or retargeted as directed by the officer in charge conducting firing. During a record practice exercise, the gun will not be retargeted before completion of the exercise except when the reduction of a stoppage makes retargeting necessary. The officer in charge of range practice will render a decision as to whether the gun will be retargeted. He will also decide whether the exercise will be continued or refired. The gun will not be targeted on a target upon which a record score is being fired.

■ 78. AMMUNITION.—Ammunition in the amounts shown in the tables for each score will be loaded in belts and inspected before firing.

■ 79. TARGETS.—The target used, its course of movement, and its speed will be as prescribed for the particular table being fired. For a complete description of targets see paragraph 86. If prepared locally, the outlines of the scoring spaces (1,000-inch targets) will be drawn so that the lines are not visible from the gun position. The targets used for the 1,000-inch firing may be placed on frames, racks, or carriages to elevate them to the height of the gun or to facilitate handling.

■ 80. PROCEDURE FOR FIRING.—*a. General.*—(1) All firing will be controlled by definite fire orders.

(2) During the initial phases of instruction firing, the officer conducting firing may at his discretion reduce the speed, targets, and the number of rounds fired from those prescribed in the tables. The object of this procedure is to place emphasis initially upon smooth, continuous tracking.

(3) A run of the target across the course once in each direction at the same speed will constitute a single score and is known as a "double run." A double run constitutes an exercise.

(4) Scores are fired in the order in which exercises are listed in the tables.

(5) During instruction firing only, when time is available, the firing of each table may be preceded by one or more practice runs.

b. Organization.—For functional purposes, an organization similar to that described in preparatory marksmanship training (par. 67) is suggested. The organization must be modified to meet the requirements of firing live ammunition.

c. Duties.—All personnel will perform the duties prescribed in paragraph 67. During instruction firing, a coach, gunner, and a loader will be at each gun. Other members may be employed in operating the range and preparing targets for firing.

d. Instruction firing.—Before firing on either the 1,000-inch or field firing range, the officer conducting firing will give a general description of the range and announce specific instructions pertaining to firing procedure.

(1) When firing at 1,000-inch moving targets two tanks are placed on the firing line to the right and left of the center stake and as close together as possible, with the horizontal rotor of the gun 1,000 inches from the targets for the parallel-level course. Another gun may be added for practice shooting if desired. They are numbered from left to right.

(2) (*a*) The coach (assistant scorer in record firing), gunner, and loader then take position at the gun. The gunner inspects the firing, elevating, and traversing mechanism and indicates that he is ready. The coach transmits this information to the officer conducting the firing.

(*b*) When all gunners are ready, the officer conducting the firing, having previously indicated to the timekeeper at the drum the course to be traveled by the target and the time of exposure for the run, commands: LOAD.

(*c*) At this command the gunner points the gun in the direction from which he expects the target to appear. The loader fully loads the machine gun and calls "Up." The gunner calls "Ready" when ready to commence the exercise. The coach signals "Ready" to the officer conducting the firing.

(*d*) When all gunners are ready the officer conducting the firing gives a signal to start the target and at the same time commands: 1. LEFT (RIGHT) FRONT, RANGE 500, ZERO (ONE, TWO) LEAD(S), SILHOUETTE NO. 1 (2), 2. COMMENCE FIRING.

(3) As the phrase LEFT (RIGHT) FRONT of the fire order is given, the gunner will start traversing toward the left (right)

of the range. Upon its appearance from behind the screen, he will engage the target, using the announced lead and firing the number of rounds contained in the belt.

(4) Immediately upon indication that all guns are ready and without further oral orders, the officer conducting firing will cause the target to be started upon its return run. The gunner engages the target as indicated in (3) above, completing his score. While the target is obscured between runs of a score, the gunner will continue to aim at the place where the target disappeared, prepared to reengage it the instant it reappears.

(5) Upon the completion of a score, guns are cleared, targets are brought to the gun positions and the results recorded, analyzed, and discussed.

(6) Upon orders from the officer conducting firing, targets will be replaced by new ones.

■ 81. RULES AND PROCEDURE FOR RECORD FIRING PRACTICE.—*a.* Record firing practice will consist of firing one of the courses prescribed in section III. Except as hereinafter specifically stated, or as specifically modified, the rules for conducting instruction firing practice as set forth in this section will apply to record practice as well.

b. Each man will complete the prescribed instruction firing for the course specified prior to firing record practice.

c. Once record practice of an individual has commenced, it will be completed without interruption by any other form of firing.

d. As a rule, record firing practice will not be fired by any candidate on the same day that he fires any part of instruction firing practice. However, when the time allotted to range practice is very limited, the officer in charge of range practice may authorize instruction and record firing on the same day.

e. No organization (company or platoon) will conduct instruction firing and record firing practice simultaneously on the same 1,000-inch moving target range unit.

f. Before firing any exercise for record, the gunner will be required, and will be given a reasonable time, to check the condition of his gun, sight, and ammunition.

g. For record firing practice, only one gun (tank) will fire on each 1,000-inch moving target range unit at a time.

h. A gunner and loader only will be at each gun during record firing. The presence of any other individual except the scoring officer and loader at or near the gun while a soldier is firing or preparing to fire record practice is prohibited. During this firing, the gunner must perform all of the operations required in firing, such as laying the gun, manipulation, and firing, without any coaching or assistance to the gunner.

i. Each gunner will complete one table of the prescribed qualification course during one order at the gun.

j. The speed of the target and its time of exposure for each run are specified in section III. The decision to disregard a score because of a failure to comply with the specified times or because of faulty operation of the target rests with the scoring officer. He will require that the target be operated in such a manner that it will traverse the prescribed course of each run at a relatively uniform rate of speed throughout its entire time of exposure. A variance of 3 seconds under or over the prescribed time for any run will be permitted. If the time of exposure is less than the prescribed time by more than 3 seconds, the gunner will be required to state whether or not he wants the score to stand before he examines the target. If he chooses to fire the exercise again he will be permitted to do so, otherwise the score will be recorded as fired.

■ 82. STOPPAGES.—*a.* In record firing practice when a misfire. stoppage, or malfunction occurs, the gunner will hold up his hand and call "stoppage." Thereafter neither the gunner nor loader will touch the gun until so instructed by the scoring officer. A scoring officer will examine the gun.

b. If a misfire, stoppage, or malfunction exists which was not the fault of the gunner, the score for that run will be disregarded and the gunner will be permitted to refire that part of the exercise.

c. If the misfire, stoppage, or malfunction is manifestly the fault of the gunner due to incorrect manipulation of the gun, the gunner will not be permitted to refire the exercise. Only that part of the exercise which was completed will be scored.

d. Should a breakage occur, the gun will be repaired or a different gun (tank) will be substituted and that part of the exercise on which the stoppage occurs will be refired. Substituted guns which have been repaired may be retargeted when the scoring officer considers such action justified.

■ 83. SCORING.—*a. General.*—Any departure from the mandatory provisions of this manual will disqualify the man affected for qualification.

(1) After a man has taken his place at the gun, all shots fired by him will count as a part of that exercise except as in paragraph 82.

(2) Failure to use the prescribed aiming silhouette for an exercise or any part thereof will result in hits in the wrong horizontal row of scoring spaces. The gunner who places his hits in the wrong row of scoring spaces will not be permitted to refire the exercise and will be scored only those hits which are found in the appropriate space for the designated aiming silhouette.

(3) A hit will be scored for each bullet hole found in the correct scoring space, except that no more hits will be counted in any scoring space(s) than the number of rounds authorized to be fired for the exercise (when a single scoring space is used for an exercise) or for a run (when a separate scoring space is used for each run of an exercise).

(4) The shot holes in the target will be counted. If the number of holes exceeds the amount of ammunition authorized for the exercise, the gunner will be penalized five points for each round in excess of the allowance.

(5) During record firing, the name of the individual will be placed on his target. No person will handle the target until after it is scored except under the direct supervision of the scoring officer or his assistant.

(6) A bullet hole which touches the line of a scoring space will be counted as a hit. (The paper having been broken into a scoring line does not necessarily mean a hit.)

(7) Holes which obviously have been made by ricocheting bullets will be counted as hits. Holes made by rocks or other foreign matter will not be counted.

(8) Ammunition not fired during the time of exposure of the target for each run of an exercise will be forfeited.

(9) After the score has been recorded for each exercise, all holes in the target will be crossed with a pencil mark and covered with a paster.

b. Computation of scores.—(1) Subject to the conditions specified in *a* above, a total of five points will be counted for each hit in a correct scoring space at 1,000 inches and for each hit on the moving target at field range (table III).

(2) The following tabulation indicates the total possible score for the three authorized courses:

	Table I	Table II	Table III	Total possible
Course A	150	150	50	350

	Table IV	Table V		Total possible
Course B	100	100	----------	200

	Table VI	Table VII		Total possible
Course C	50	50	----------	100

c. Score cards.—A score card will be kept for each person firing. This card will show the scores made during record firing practice. Each individual entry for record practice will be made in ink or indelible pencil and will be authenticated by the scoring officer. Erasures are not permitted. Alterations will be made only by the company commander of the officer who acted as scorer. Such corrections will be authenticated by the officer making the correction.

■ 84. INDIVIDUAL CLASSIFICATION AND QUALIFICATION.—For individual classification and qualification scores, see AR 775–10 and current War Department circulars.

■ 85. SUSPENSION OF COACHING RESTRICTIONS DURING RECORD FIRING.—The provision of this manual which prohibits coaching during record practice is suspended during the period for which no additional compensation for arms qualification is authorized, with the following exceptions:

a. The coach will not use any mechanical aid, such as an aiming device, to assist the gunner.

b. The coach will not touch any part of the gunner's body while the gunner is sighting or firing.

■ 86. MARKSMANSHIP TARGETS.—The targets used for firing marksmanship courses are as follows:

a. 1,000-inch moving target.—(1) The target frame for all 1,000-inch firing is 3 feet 6 inches by 2 feet 6 inches. It is made of ¾- by 1½-inch lumber, halved and joined squarely at the corners. The frame is covered with target cloth to provide a bearing surface for the paper target.

(2) The 1,000-inch target is shown in figure 34. Silhouette No. 1 represents the size of a tank at approximately 300 yards, No. 2 represents 600 yards, and No. 3, 900 yards.

b. Field range 75-mm tank gun target.—The target for field range firing is a 5- by 8-foot plain, light-colored panel used when firing the preliminary and record practices of table III.

■ 87. CONSTRUCTION OF 1,000-INCH MOVING TARGET RANGE.—AR 750–10 prescribes the danger areas for target ranges. Due to the small size of the 1,000-inch range, a location can usually be found without difficulty. A level open space about 70 yards long (in the direction of fire) and about 20 yards wide is required for each moving target range unit. A single unit is necessary for each tank platoon. It will greatly facilitate the conduct of 1,000-inch firing to have two of these range units per platoon. A range unit consists of two runways in which a sled target moves to simulate the various directions of movement and speeds of probable combat targets. Movement of the sled is actuated by a wire cable which runs from a hand-operated drum through a system of pulleys and is fastened to both ends of the sled.

a. The dimensions and plan of construction of the two runways, that is, the parallel-level and parallel-hilly courses, are given in figure 35.

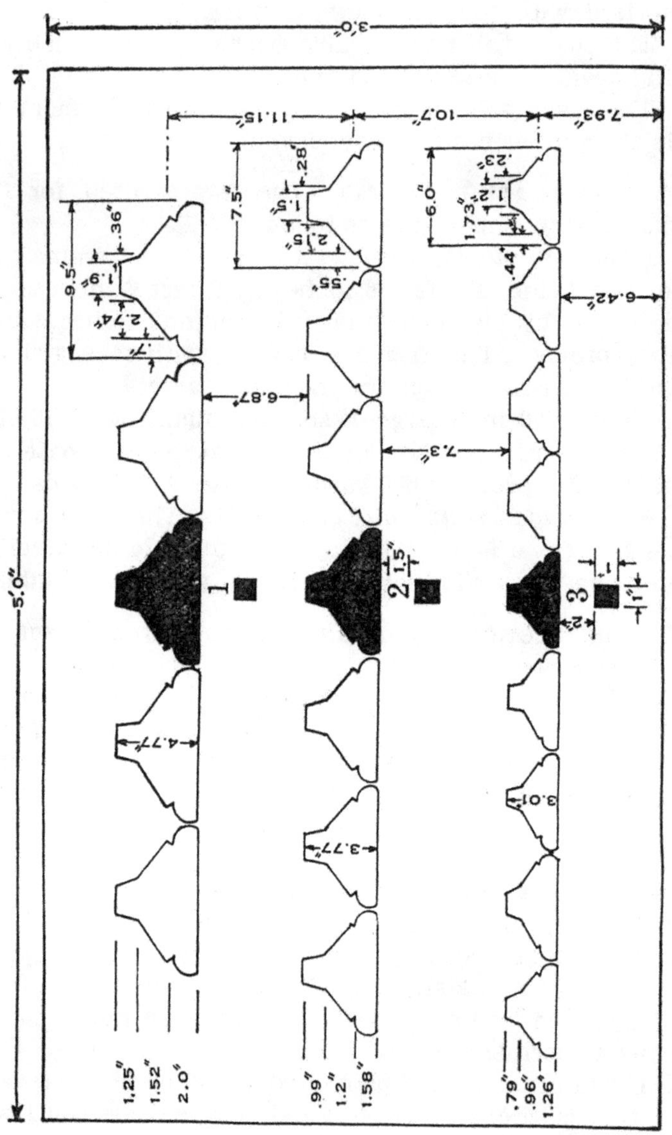

FIGURE 34.—1,000-inch moving target.

b. Figure 36 is a sketch of the arrangement of the whole unit showing the positions of the pulleys and the hand-operated drum (reel).

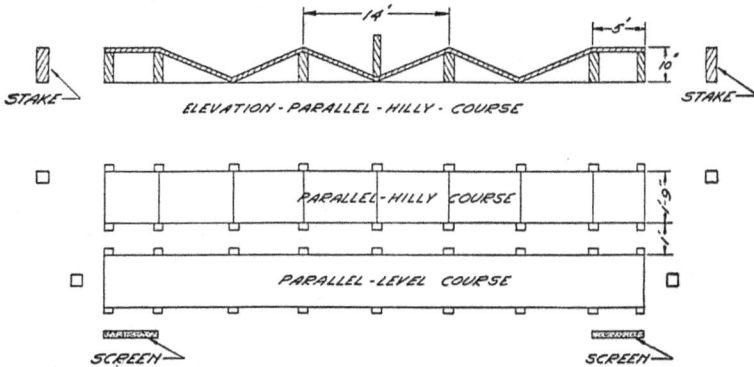

FIGURE 35.—Dimensions and plan of construction, 1,000-inch moving target range.

c. Figure 37 shows the hand-operated drum. A commercial wooden cable drum was used to make this reel. The weight and relatively large diameter of the drum tend to resist sudden

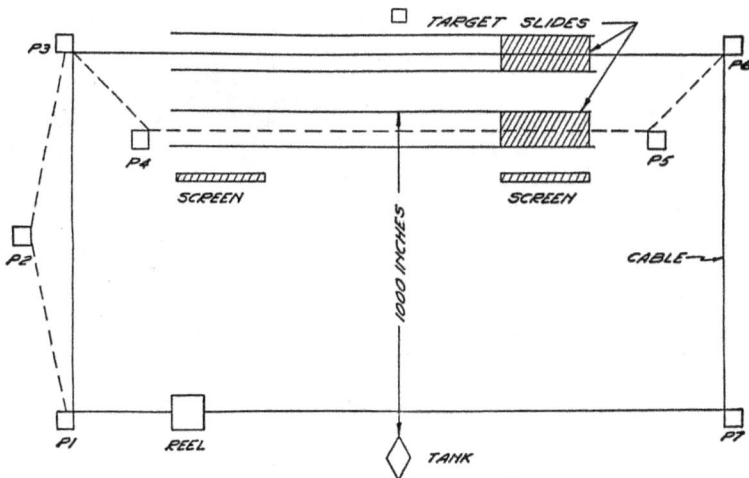

FIGURE 36.—Arrangement of 1,000-inch moving target range unit, showing positions of pulleys and hand-operated drum (reel).

changes in speed, thus insuring smooth operation of the sled. Two 2½-inch swivel eye pulleys are attached to the base of the drum mount to bring the running cable close to the ground.

d. The offset pulley P2 is a snatch pulley, so placed that when the sled is used on the parallel-hilly course, the wire may be run around one of them as shown by the dotted lines in figure 36 to take up the slack due to the difference in length of each course.

e. For the parallel-level course, the wire runs from the reel through P1, P3 to sled then P6, P7 to the reel. For the

FIGURE 37.—Hand-operated drum (reel), 1,000-inch moving target range.

parallel course the circuit is: reel P1, P2, P3 to the sled, P5, P6, P7, and reel.

■ 88. TARGET SLED.—The sled in which the 1,000-inch moving targets are mounted should be heavy lumber in order to provide a low center of gravity which is necessary for smooth operation of the sled. The target holder can be made of ¾-inch material. The sled should be mounted on casters.

The standard which is attached to the top of the sled should be made to hold the target firmly and still allow it to be withdrawn easily.   A piece of twisted wire cable should be fast-

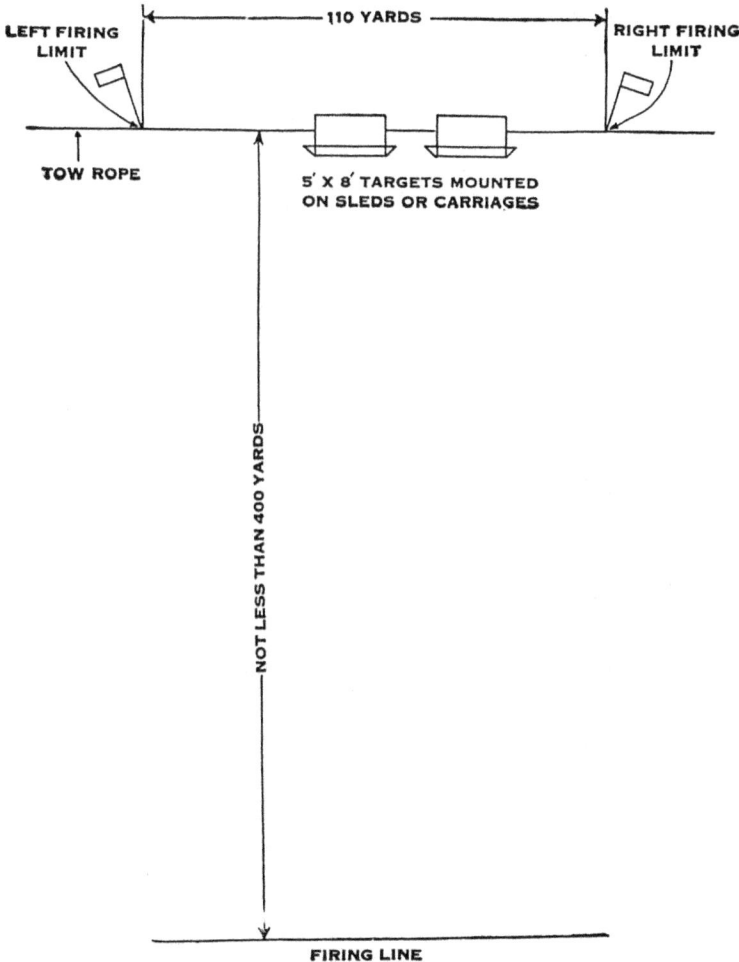

FIGURE 38.—Moving target, combat range.

ened across each end of the sled to provide a means for attaching the snaps of the towing cable.

■ 89. CONSTRUCTION FOR CALIBER .30 OR 75-MM KNOWN RANGE MOVING TARGET FIRING.—a. An approximately level or gently

111

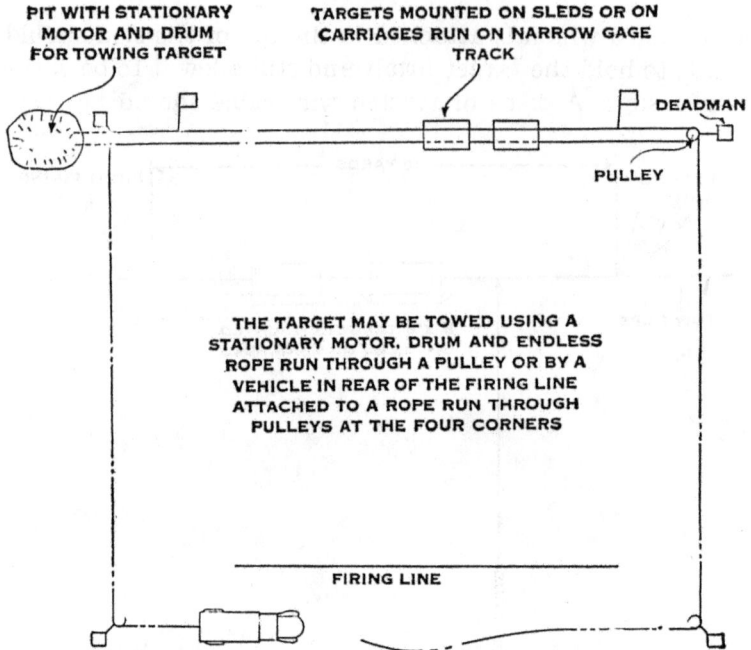

FIGURE 39.—Range for caliber .30 or 75-mm moving field target.

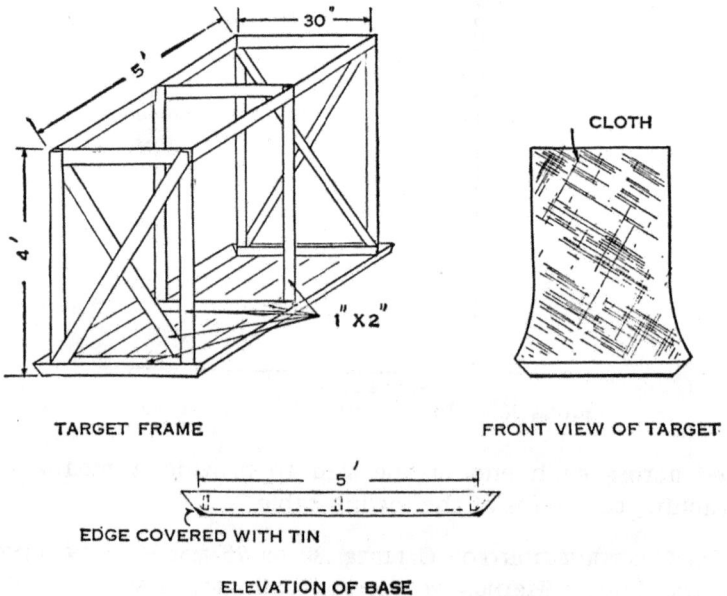

TARGET FRAME

FRONT VIEW OF TARGET

EDGE COVERED WITH TIN

ELEVATION OF BASE

FIGURE 40.—Target frame and sled for towed-target range.

rising piece of ground should be selected for this range. It should be at least 1,000 yards in depth and of sufficient length to permit convenient operation of the moving target. Figures 38 and 39 show a moving target range, including two methods of towing the target. The construction of pits for the scoring and target details will facilitate scoring and changing targets. It is desirable also to have telephone communication between the pits and the firing line.

b. The set-up shown in figures 38 and 39 is flexible; that is, the length of the target run and the distance between firing line and target can be varied as desired. By placing the gun in successive firing positions, ranges between 200 and 500 yards may be obtained. Flags are placed as shown in the figures to indicate the right and left firing limit. These limits will vary with the distance from gun to target. They must conform to the conditions prescribed in AR 750–10. The target used when firing on this range is a 5- by 8-foot plain, light-colored panel. The panel is mounted, long edge horizontal, on a suitable sled or carriage as indicated in figure 40.

CHAPTER 4

TECHNIQUE OF FIRE

SECTION I

GENERAL

■ 90. DEFINITION.—The application of effective fire on the target is called technique of fire.

■ 91. SCOPE.—The technique of fire as discussed in this chapter includes—

Characteristics of fire.
Terrain features.
Targets.
Ammunition.
Firing positions.
Range determination.
Estimation of speed.
Target designation.
Fire distribution.
Fire control and adjustment.
Fire orders.

SECTION II

CHARACTERISTICS OF FIRE

■ 92. RANGE OF 75-MM GUN.—The effective range of the 75-mm tank gun M2 is beyond the range at which the gunner can effectively aim on the target. Most targets will be engaged at ranges below 1,000 yards. However, targets may be engaged at ranges as great as 2,400 or 3,000 yards.

■ 93. CLASSES OF FIRE.—a. At present the gun is used only in direct laying. It may later be equipped to fire by indirect laying methods.

b. The gun may be used in direct assault when the medium tank platoon is part of the leading wave of attack. It may also be used to support from defiladed positions the attack of other tanks both light and medium. In the support role fire is delivered between tanks of the leading wave, overhead fire when elevated firing positions can be obtained, and on targets on the flanks of the supported unit.

SECTION III

TERRAIN, TARGETS, AMMUNITION, FIRING POSITION

■ 94. GENERAL.—This manual does not teach tactical employment except insofar as is necessary to move tanks into position to fire on field targets. However, before engaging in combat practice firing, the gunner must know and be able readily to recognize common terrain features, understand where to look for targets, and know where to place the tank when firing from the stationary position.

■ 95. TERRAIN.—To teach the gunners the names of terrain features and where to look for targets, select a varied piece of terrain and place various targets in position as they would probably be found in combat. From an observation point, indicate the names of the various terrain features to the assembled class. Next walk through the area pointing out the various features and the targets.

■ 96. TARGETS.—*a.* Targets should be engaged first in the order of danger to the tank itself and next their danger to supported troops. When supporting other tanks, targets are engaged in accordance with their danger to the supported tanks. The order of engagement is—

(1) Armored vehicles, the armament of which is effective against our own tanks.

(2) Antitank guns, including field artillery.

(3) Targets the reduction of which will materially aid the maneuver of our own forces such as—

Lightly armed and armored vehicles.

Machine guns.

Personnel.

b. The supply of 75-mm ammunition in the tank is limited. It is therefore essential that it not be wasted on unimportant targets. Do not use the 75-mm gun when the machine gun will do just as well.

■ 97. AMMUNITION.—*a.* Three types of ammunition are to be furnished for tank firing: high-explosive, armor-piercing, and smoke. For description of this ammunition except smoke, see chapter 1. In general the weapons and projectiles to be used for various classes of targets are as follows:

Target	Weapon and projectile
Personnel	Machine gun 37-mm cannister 37-mm shell 75-mm shell
Machine guns, trucks, unarmored vehicles, unarmored antitank guns	37-mm HE or AP 75-mm HE
Armored vehicles	37-mm AP
Armored antitank guns, emplacements	75-mm AP

b. The tabulation above must not be followed blindly. Do not use 75-mm shell against a target when the machine guns in the tank or the 37-mm gun will be just as effective. On the other hand, do not hesitate to fire 75-mm high-explosive ammunition against armored vehicles when armor-piercing ammunition is not available. A hit by a 75-mm shell on a track will disable the tank.

c. Smoke shell may be used for certain operations. Smoke will be used to blind distant targets that cannot be reached by a mortar and to mark targets for air bombardment.

d. The 75-mm armor-piercing projectile must actually hit the target to be effective. The 75-mm shell bursts upon impact and will cover effectively an area 5 yards in depth by 30 yards in width. Large fragments of the shell are effective up to 150 yards from the point of burst.

■ 98. POSITION.—*a.* Although some moving tank firing may be done, most of the firing from the 75-mm tank gun will be from a stationary position. As the tank is far more vulnerable when stationary than when moving, it is necessary that firing positions be selected with a view to concealment and cover. When concealment is not available, the tank should halt only long enough for the gunner to fire one or two shots. It should then move forward rapidly.

b. For supporting the attack of other tanks, concealed or covered positions are always selected, when available. Such positions may be in the edge of woods, high brush, or other vegetation that will conceal a tank. It may be on the reverse slope of a hill with only enough of the hull of the tank visible to permit direct aiming of the gun. Avoid positions where the tank will be silhouetted against the skyline or against a sharply contrasting background. A position with the sun behind the tank is excellent as the rays of the sun will somewhat blind the enemy. *However, do not silhouette the tank against the sky.*

c. In training, in selection of firing positions, conduct the class, by walking, through an area and show them good and bad positions. Place tanks, some in proper and some in improper positions, cause the class to view them from a suitable observation point, and point out errors in the selection of positions.

SECTION IV

ESTIMATION OF RANGE AND SPEED

■ 99. GENERAL.—The rapidity with which armored vehicles move prohibits any slow process of determination of range and speed. For this reason, the lead table given in paragraph 67 was devised. Range and speed must often be estimated while traversing the gun to engage the target. View from the tank is limited and its constant motion further impairs observation. Therefore, it is essential that men learn to estimate key ranges and speeds, fire with the estimated data, and correct by observation.

■ 100. RANGE ESTIMATION.—There are two methods of range determination: estimation by eye and firing the gun. The more usual method for the 75-mm gunner will be estimate by eye.

a. Estimation by eye.—Select a varied piece of terrain. Place tanks, half-track and other vehicles, antitank guns, and other targets at 300, 600, 900, 1,200, and 1,500 yards, if possible,

117

from an observation point. Assemble the class and explain that when estimating range by eye—

  (1) Targets appear nearer—

    (a) When the object is in a bright light.

    (b) When the color of the object contrasts sharply with the color of the background.

    (c) When looking over water, snow, or a uniform surface such as a wheat field.

    (d) When looking downward from a height; in the clear atmosphere of high altitudes.

    (e) When looking over a depression most of which is hidden.

  (2) Targets appear distant—

    (c) When looking over a depression most of which is visible.

    (b) When there is poor light or fog.

    (c) When only a small part of the object can be seen.

    (d) When looking from low ground upward toward higher ground.

Cause the men to become familiar with the appearance of targets at the key ranges under as many of the conditions given above as practicable. Make sure that they have firmly fixed in mind the key ranges of 300, 600, and 900 yards. Mount the men in tanks and have them view the key ranges through the sights and vision devices. For ranges greater than 900 yards, select what is considered the midpoint between the gun and target, estimate the range to this point and double it.

  b. Firing a gun.—This method has little application to the 75-mm gun. A gun, usually a machine gun, is fired at the target with an estimated range, the strike is rolled into the target, the range determined by the sight, and announced to the gunner.

■ 101. Estimation of Speed.—Speed is estimated as slow (below 10 mph), medium (10–20 mph), and fast (above 20 mph). On a piece of terrain (par. 100), assemble the class at the observation point and explain to them that they are to estimate in these three speed ranges only and *not try to estimate speed to the exact number of miles per hour.* At 300, 600, and

900 yards from the observation point have tanks driven across the range at right angles to the line of observation. Have them first driven at speeds below 10 miles per hour, next at speeds between 10 and 20 miles per hour, and finally at speeds above 20 miles per hour. Announce to the class at what speed, slow, medium, or fast, the tanks are being driven. Next have the tanks drive at unannounced speeds and require the class to estimate the speed and range and to announce the lead to be used.

SECTION V

TARGET DESIGNATION

■ 102. GENERAL.—Two methods are used to designate targets: oral designation and firing a gun.

■ 103. ORAL DESIGNATION.—*a. Elements.*—In this method the target is designated by word of mouth with or without mechanical aids such as interphones and radio. Such designation is assisted where practicable by pointing. The three elements of oral target designation are: range, direction, and description.

(1) *Range* is announced by the word "Range" followed by a statement of yardage such as "600." The inclusion of the word "yards" is unnecessary and is omitted.

(2) *Direction* is indicated orally and is supplemented by pointing if practicable. Figure 41 shows a method of orally designating direction. For instance, direction to target No. 3 should be designated as "Right flank."

When the target is indistinct and there is a distinct terrain feature nearby, direction may be indicated by using the terrain feature as a reference point. Example: Target No. 10 (fig. 41) may be designated by—

RANGE 800, FRONT, SMALL BUSH, LEFT TEN MILS (or so many yards), MACHINE GUN.

(3) *Description* consists of a word or two to describe the target such as: machine gun, troops, AT gun, tank. When the target is distinct, description may be omitted.

119

b. Complete designation.—The following are examples of target designation (see fig. 41):

(1) *Target No. 8.*

RANGE 900, LEFT FRONT, SMALL HILL, AT GUN, ON CREST.

(2) *Target No. 6.*

RANGE 600, LEFT REAR, FENCE CORNER, MACHINE GUN.

(3) *Target No. 2.*

RANGE 700, RIGHT FRONT, TANK.

■ 104. FIRING THE PIECE.—*a.* In this method the person designating the target says: RANGE 800 (or as estimated),

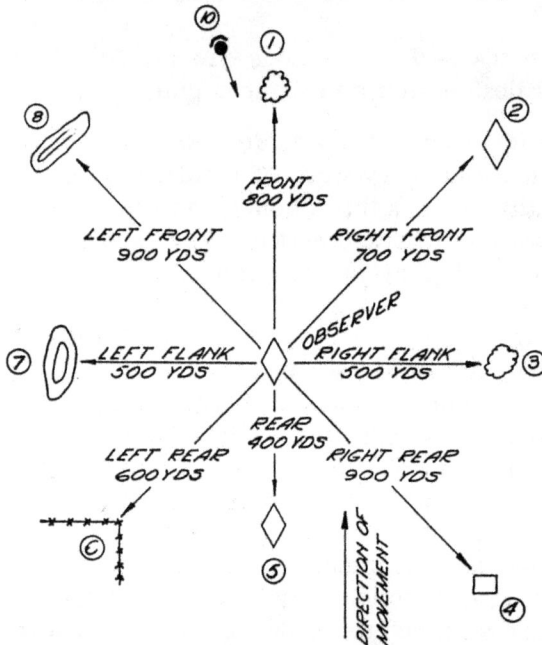

FIGURE 41.—Direction for oral target designation. "Front" is the direction in which the observer's tank is moving.

FRONT (or other direction), WATCH MY BURST. He then lays on the target with the machine gun, fires and rolls the strike into the target.

b. This method is particularly applicable for the tank commander to designate targets to other tanks of his platoon using radio. The tank driver may also use the sponson or bow guns as the case may be to designate targets he has observed. However, as these guns are fixed as to direction and the sponson guns on light tanks fixed as to elevation, he must maneuver the tank to bring fire on the target.

■ 105. SIGNALS.—The following signals may be used in target designation:

Attention_____ Grasp a portion of the gunner's clothing and jerk him.

Range:

100–400 yards__ Expose one finger to the gunner for each 100 yards of range including 400 yards.
yards.

500 yards_____ Expose fist.

Above 500 yards____ Expose fist for each multiple of 500 yards and one finger for each 100 yards above this range. Example: 700 yards—expose closed fist and follow with exposure of two fingers.

■ 106. USE OF RADIO.—Voice radio is used by the platoon leader to designate targets for individual tanks of his platoon. No complicated code should be used, the platoon leader may merely say "Watch my burst" and fire the caliber .30 machine gun at the target to be designated. Example: TANK NO. 5— RANGE 700, WATCH MY BURST.

■ 107. TRAINING.—Select a varied piece of terrain. Set up various targets, antitank guns, machine guns, and others, in suitable positions. Walk the class through the area and point out the target. Next assemble the class at the observation point and have them designate the various targets. The men are then mounted in tanks, driven through the area, and designate targets while the tank is moving. The use of tanks and other vehicles moving at various ranges and speeds should be included in this instruction. Men will be required to designate such targets and to announce the lead.

BASIC FIELD MANUAL

SECTION VI

FIRE DISTRIBUTION, ADJUSTMENT, CONTROL, AND
ORDERS

■ 108. FIRE DISTRIBUTION.—*a.* Fire distribution as it applies
to the 75-mm tank gun implies the engagement of targets
in their order of importance and the use of all guns of the
platoon or section to engage several targets effectively and

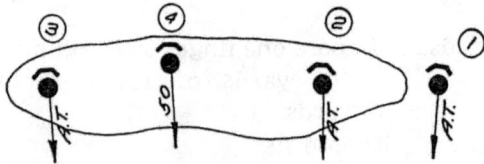

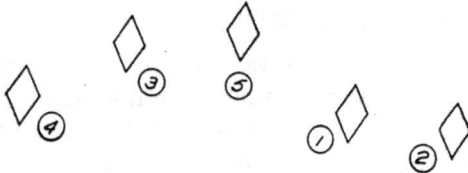

FIGURE 42.—Fire application, tank platoon.

simultaneously. Distribution of fire of the caliber .30 machine guns mounted in tanks and the 37-mm gun is covered
in FM 23–50 and 23–81.

b. A single tank engages targets most dangerous to itself.
The section or platoon distributes its fire so as to engage
targets most dangerous to the section or platoon. In
platoon or section formation each tank commander is responsible for targets immediately to his front. In addition,

commanders of flank tanks are responsible for targets on their flank. Each tank commander observes to his right and left and immediately brings fire to bear on targets in front of other tanks when he is not engaged on his own front. The tank commander must not hesitate to switch fire to a more dangerous target. Figures 42 and 43 show schematically the application of fire under certain assumed conditions.

(1) In figure 42, the platoon is advancing in line. Each tank is responsible for the area to its immediate front. However, gunners do not hesitate to fire on targets in front of other tanks when they are not themselves engaged with more important targets. In the situation in figure 42, an antitank gun at (1) opens fire. Both tanks Nos. 1 and 2 open fire. Tank No. 3 opens fire on machine guns at No. 4. The antitank gun at (3) opens fire. Tank No. 3 immediately shifts fire to antitank gun at (3), a more dangerous target than the machine guns. An antitank gun at (2) opens fire and tank No. 1 switches fire to that target. The platoon leader, tank No. 5, controls the fire of the others.

(2) Figure 43 shows tanks attacking with one section covering the target while the other moves to attack.

c. (1) In tank versus tank combat, the platoon commander, when practicable, places one section in concealment or defilade to fire on the enemy tanks while with the remainder of the platoon he maneuvers to place himself in the flank of the enemy. Figures 43 and 44 show diagrammatically methods of fire distribution when hostile tanks are engaged.

(2) The hostile tank most dangerous to the platoon, section, or individual tank must be engaged first. This does not necessarily mean the closest target, as a stationary tank within range is more dangerous than a moving tank. In figure 45 three hostile tanks are approaching. One of these tanks stops, the others continue on toward your tank. Halt and fire on the halted tank. It is more dangerous as its fire will be more accurate than the fire of the moving tanks. Furthermore, fire will be more accurate against the stationary tank than against the moving tanks.

d. The following points should be emphasized:

(1) Fire on the target most dangerous to you or your unit.

(2) Do not hesitate to switch your fire to a more dangerous target.

(3) Seek cover or concealment from which you may fire halted. If such cover is not available halt for a few seconds, fire and then advance.

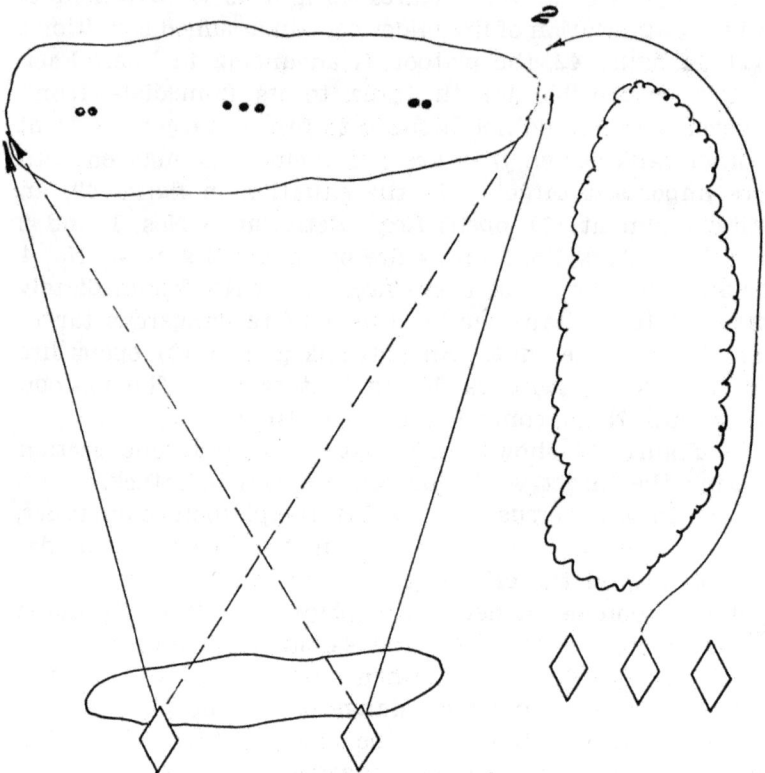

FIGURE 43.—Fire distribution, tank platoon in attack showing maneuver.

(4) When under fire advance by zigzagging.

(5) Use fire and movement.

■ 109. FIRE ADJUSTMENT.—*a. General.*—(1) Fire adjustment is one of the most important factors in any firing. It must be particularly rapid and accurate in tank firing. Tanks can halt in one position for only a short time to fire as the muzzle blast will quickly disclose even a well-concealed position.

124

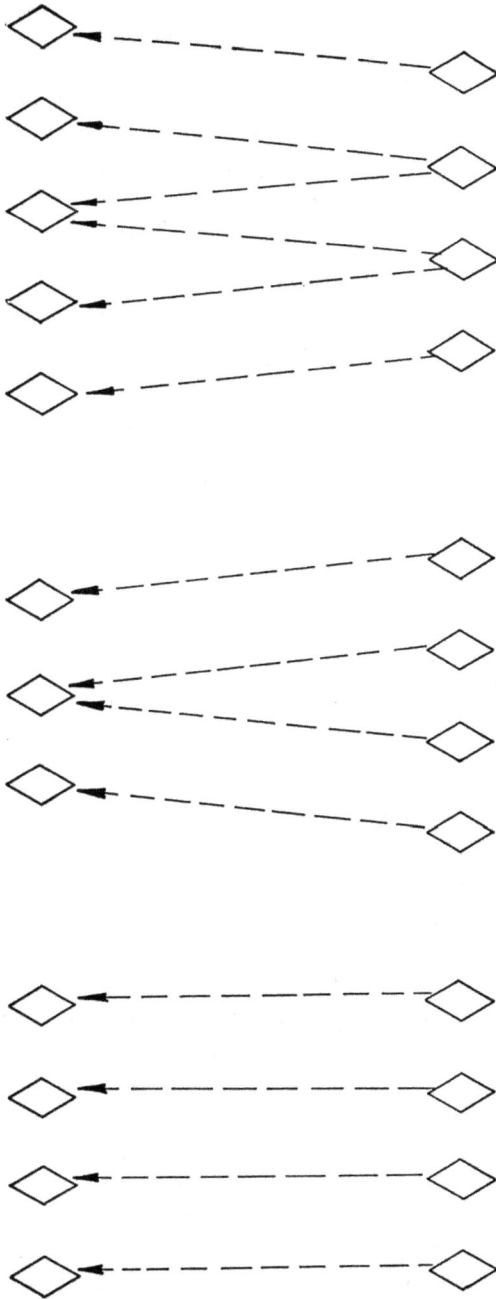

③ Tanks engaging superior number of hostile tanks.

② Tanks engaging smaller number of hostile tanks.

① Tanks engage tank opposite them in hostile formation.

FIGURE 44.—Fire distribution, tank versus tank combat.

(2) Fire adjustment falls into two categories: adjustment on moving targets and adjustment on stationary targets.

b. Traversing and elevating mechanism.—Before proceeding to training in fire adjustment, the gunner must be taught how to use the elevating and traversing mechanism on the gun to make such adjustments. Although, he has been taught manipulation in the exercises on the 1,000-inch range, he must, before engaging in combat practice firing, know how

TANK HALTED

HALT YOUR TANK AND
FIRE IN THE STATIONARY
TANK.

FIGURE 45.—Engaging enemy tanks.

much to move each handwheel to make corrections in range and traverse.

(1) *Traversing.*—The gun has a total traverse of 28° by use of the traversing handwheel. Additional traverse is made by turning the tank. As the gunner nears the limit of traverse, either right or left, he must signal the driver to turn the tank so that the gun may be laid on the target. One turn of the traversing handwheel moves the point of aim 1°6′ or about 6 yards at 300, 12 at 600, 18 at 900, 24 at 1,200, and 30 at 1,500

yards. The traversing handwheel therefore allows for a fine adjustment on the target. An observer in calling corrections may call the number of yards right or left to move the point of aim. The gunner then lays back on the target and traverses so many turns to the right or left. When designating targets by reference point, the observer gives the distance as so many yards right or left of the reference point. The gunner lays on the reference point and traverses right or left the number of turns necessary. For example, a target is designated as range medium, reference point lone tree to the front, 50 yards right, antitank gun. The gunner lays on the reference point and traverses right or left the number of turns necessary.

(2) *Elevating.*—One turn of the elevating handwheel raises or lowers the point of aim 1°8'. This amounts to a change in range of approximately 1,500 yards. Thus a slight movement of the elevating handwheel will make quite a change in range. A movement of one-fifteenth of a turn of the handwheel making a change of 100 yards. The gunner must be particularly careful with the manipulation of this handwheel.

c. Adjustment on moving targets.—The gunner lays on the target with estimated range and lead and fires. At the instant of firing he must make and retain a mental picture of the target in relation to the sight. He keeps his eyes on the target and watches the tracer of the round or for the strike, he immediately relays on the target and increases or decreases elevation and lead in accordance with estimated corrections. Figure 46 shows adjustments to be made to bring the strike on the target.

d. Fire adjustment on stationary targets.—(1) *Ranges 1,000 yards or below.*—Lay on the target with the estimated range, fire, and observe the tracer or strike. Make estimated adjustments in aiming and fire again.

(2) *Ranges greater than 1,000 yards.*—Targets will usually be indistinct at such ranges and great care must be exercised in laying. The tank commander must assist the gunner in sensing his shots. Lay on the target with the estimated range and fire. If the shot is over, aim with the next lower numbered range line. If it is short, aim with the next higher range line. Fire and proceed as above until both an over and a short is obtained. Aim again with a range half-way

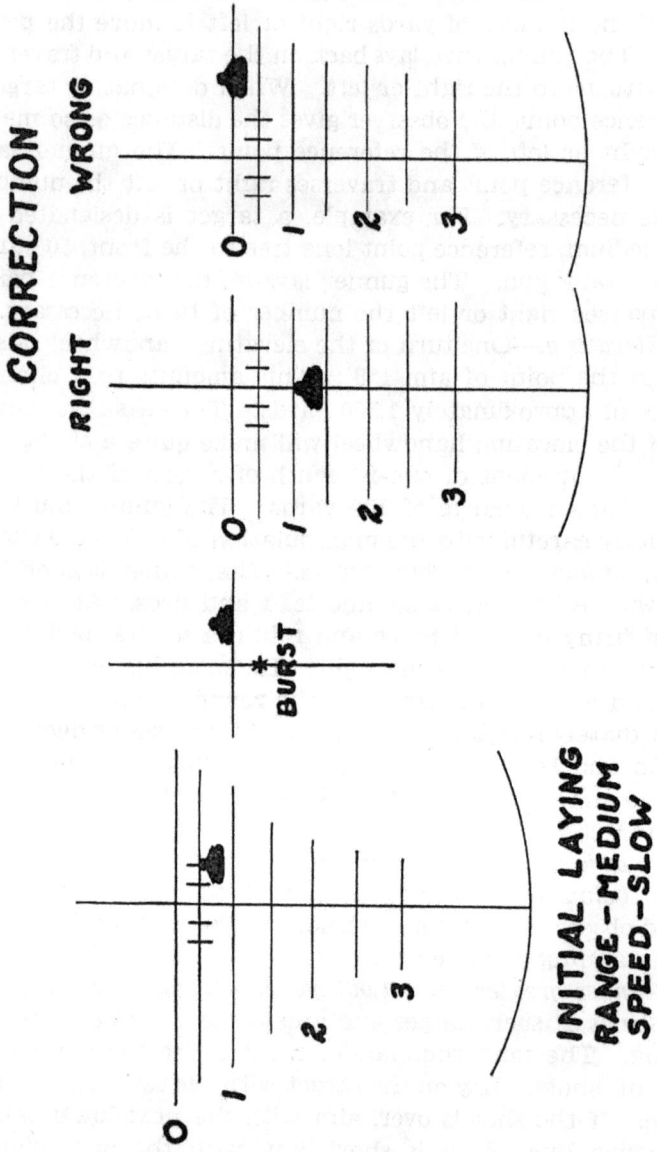

FIGURE 46.—Adjustment of fire on moving target.

between the over and the short ranges and fire. Continue this adjustment until a hit is obtained.

■ 110. FIRE CONTROL.—*a.* Fire control implies the ability of the leader to open fire at the instant he desires, adjust the fire of his guns upon the target, switch fire to other targets, and cease firing at will. The noise of the tank makes communication difficult. Commands must be brief and signals are used to a great extent. The tank is equipped with an interphone system.

b. Fire control for initial attack.—When practicable, the platoon leader points out to the tank commander the initial targets for the attack or designates areas of responsibility for each tank. When a portion of the platoon is to cover the movement of the rest of the platoon, the platoon leader may designate the time for opening fire or prescribe that fire be opened on signal.

c. During the attack the platoon leader controls fire by radio and flag signal. Much must be left to the initiative of the individual tank commander and gunner. The platoon leader exercises tactical control, but the locating of targets and opening fire on them are primarily functions of the tank commander. The success of an attack will depend upon the alertness of the tank crew.

■ 111. FIRE ORDERS.—*a.* A fire order for tank firing must be brief. It usually consists of—

     Range.
     Direction.
     Description.
     Lead (moving targets only).
     Time of opening fire.

All the above elements are not necessary. For example: The tank commander might say: FRONT, RANGE 600, TANK, COMMENCE FIRING. The gunner lays on his target as previously described, estimates the lead, and fires.

b. Examples of fire orders:

     RANGE 600, RIGHT FRONT, TANK, TWO LEADS, COMMENCE
               FIRING

RANGE 800, FRONT, MACHINE GUN, NEAR DEAD TREE, COMMENCE
FIRING

c. The tank commander must designate the gun he wishes to use on the target and may designate the projectile. An example of such an order is—

RIGHT FRONT, RANGE 600, ENEMY TRUCK, 75-MM GUN, HE, COMMENCE FIRING

In the example above, the tank commander might designate both the 75-mm gun and 37-mm gun to fire on the target.

d. In most cases the type of projectile will not be designated, as the gunners are taught to use the proper type habitually.

CHAPTER 5

COMBAT PRACTICE FIRING

■ 112. GENERAL.—*a.* Combat practice gives the gunner practice in firing at field targets under simulated combat conditions. The effectiveness of instruction depends upon thorough preparation and the initiative of the officer in charge. Exercises must be made interesting or little value will be received from the instruction.

b. Each exercise is a combat problem in which the platoon leader, section leader, or tank commander is given briefly the enemy situation, friendly situation, and mission. Do not give in detail a general situation. It is sufficient for the leader to know the mission of his unit, the scheme of maneuver, and fire support.

c. The amount of ammunition available will be small. Early exercises will be conducted with 37-mm ammunition and later exercises with the 75-mm.

d. Ranges should provide, when practicable, an open space of approximately 2,000 yards. The range plans shown in FM 23–81 are suitable for 75-mm gun-firing.

e. Exercises are held for individual tanks, section, and platoons. They are designed to give practice in—

    Fire adjustment
    Support of an attack
    Attack by a section or platoon

f. Fire and movement is stressed. Particular attention must be given to selection of firing positions.

■ 113. FIRST EXERCISE—FIRE ADJUSTMENT, STATIONARY TARGET.—Place screen targets at approximately 600 and 900 yards from the gun position. Have tank moved into a defiladed position. Require the tank commander to designate the target and the gunner to fire on the target to adjust fire. Use 37-mm ammunition. Allow not more than five rounds per exercise.

■ 114. SECOND EXERCISE—FIRE ADJUSTMENT ON MOVING TARGET.—Use a moving target range as described in FM 23–81.

Place the tank in defilade between 600 and 900 yards from the starting point of the target. Cause the target to move and require the gunner to adjust fire on it. Speed should be medium. Allow not more than five rounds of 37-mm ammunition.

■ 115. SUPPORT OF A TANK ATTACK.—*a.* Use both stationary and moving targets at ranges of 500 to 1500 yards from the initial firing positions. Assemble the tank section or platoon under cover in rear of first firing positions.

b. From an observation point give the situation to the leader, telling him that he is to support by fire the advance of another tank unit. Point out one or two targets. Allow the leader if he desires, to bring his tank commanders to the observation point. Give time for selection of routes and positions, then require the platoon to move into position to support the attack. Use disappearing and moving targets after the tanks are in firing position.

■ 116. ATTACK BY A PLATOON OR SECTION.—Set up targets as in paragraph 115. From an observation point give the situation to the leader. Point out any safety precautions necessary. Require the leader to conduct his platoon to attack in the zone assigned. Tanks to be halted upon arrival at a line marked by flags. As the platoon advances cause various disappearing targets to appear and start moving targets.

■ 117. SAFETY PRECAUTIONS.—*a.* Range areas and safety precautions must conform with instruction set forth in AR 750–10.

b. Misfires are handled as explained in paragraph 23.

c. Markers will be placed so as to clearly define right and left limits of fire.

d. The starting point of the vehicle towing targets must be a safe distance on the flank opposite that on which the target appears.

e. Vehicles and personnel working on the course will be equipped with red flags and must be directed by definite signals and commands.

f. Each tank commander will see that the gunner—

(1) Never endangers the target detail.
(2) Never fires outside the prescribed safety limits.
(3) Ceases firing upon command.
(4) Clears his gun when ordered.
(5) Does not load until the range is clear.

INDEX

○

Also Now Available!

FM 17-68
WAR DEPARTMENT FIELD MANUAL

M5 STUART LIGHT TANK
CREW MANUAL
-1944-

BY WAR DEPARTMENT

PERISCOPE FILM LLC

Visit us at:

www.PeriscopeFilm.com

TM 9-759
WAR DEPARTMENT MANUAL

M4 SHERMAN
MEDIUM TANK
TECHNICAL MANUAL

BY WAR DEPARTMENT

PERISCOPE FILM LLC

FM 17-76
WAR DEPARTMENT FIELD MANUAL

M4 SHERMAN
MEDIUM TANK
CREW MANUAL

BY WAR DEPARTMENT

PERISCOPE FILM LLC

©2013 Periscope Film LLC
All Rights Reserved
ISBN#978-1-937684-49-5
www.PeriscopeFilm.com

www.ingramcontent.com/pod-product-compliance
Lightning Source LLC
Chambersburg PA
CBHW052104090426
42741CB00009B/1671

9 781937 684495